T0197822

Proven Climate Solutions

Proven Climate Solutions

Leading Voices on How
to Accelerate Change

Edited by BF Nagy

ROWMAN & LITTLEFIELD
Lanham • Boulder • New York • London

Published by Rowman & Littlefield
An imprint of The Rowman & Littlefield Publishing Group, Inc.
4501 Forbes Boulevard, Suite 200, Lanham, Maryland 20706
www.rowman.com

86-90 Paul Street, London EC2A 4NE

Copyright © 2024 by BF Nagy

All rights reserved. No part of this book may be reproduced in any form or by
any electronic or mechanical means, including information storage and retrieval
systems, without written permission from the publisher, except by a reviewer who
may quote passages in a review.

British Library Cataloguing in Publication Information Available

Library of Congress Cataloging-in-Publication Data

Names: Nagy, B. F., 1956– editor. | McKibben, Bill.
Title: Proven climate solutions : leading voices on how to accelerate
  change / edited by BF Nagy.
Description: Lanham : Rowman & Littlefield, [2024] | Includes
  bibliographical references and index.
Identifiers: LCCN 2023049474 (print) | LCCN 2023049475 (ebook) | ISBN
  9781538186534 (cloth) | ISBN 9781538186541 (epub)
Subjects: LCSH: Climate change mitigation. | Environmental protection.
Classification: LCC TD171.75 .P76 2024 (print) | LCC TD171.75 (ebook) |
  DDC 363.7/06—dc23/eng/20240307
LC record available at https://lccn.loc.gov/2023049474
LC ebook record available at https://lccn.loc.gov/2023049475

♾️™ The paper used in this publication meets the minimum requirements of
American National Standard for Information Sciences—Permanence of Paper
for Printed Library Materials, ANSI/NISO Z39.48-1992.

For Elizabeth Nagy Loan
who has never once stopped believing in what is right.

Contents

Foreword

Mary D. Nichols

If you are worried, angry, depressed about the global warming crisis, whether you are a student, scientist, or engineer, an ecologist, activist, or any sort of person who is aware of the rapidly changing climate, you will find much to engage you here.

And if, like me, you sometimes are hard-pressed to feel hopeful that those with the power to change will do so in time to avoid the total erasure of places and societies on this planet, you will be heartened to know that a range of experts from various fields have sensible, practical solutions that are already here or could be put in place by businesses and governments if we push them to act now.

We used to hear we have a decade or less to stop polluting the atmosphere with greenhouse gases. The evidence of growing damage to oceans, fresh water, plant and animal life, and human health caused by rising temperatures is everywhere. Mainstream news organizations have moved from ignoring the issue to breathless coverage of fires and floods. Some are even turning to scientists to help explain the science behind the atrocities.

Not surprisingly, humans react differently to the existential threat. More and more people in democratic societies are moved to action. Marches, lawsuits, and voluntary efforts to reduce fossil fuel consumption are rising. Inventors, entrepreneurs, and investors are creating and refining technologies that can make real, measurable reductions in the growth of emissions. But the upward curve of planet-harming emissions is not bending yet. The giant corporations and nationally owned petroleum producers around the world are increasing, not reducing their production, and their profits have reached record highs. And with the rise of authoritarian, radically right-wing governments, countries that signed the Paris Accords agreeing to cap and reduce their emissions (every country on the planet except for the

United States under the former administration) are now stalling or quietly reneging on their commitments.

Breaking the grip of fossil fuel dependence and halting the destruction of the forests, oceans, and lands that store so much carbon dioxide, without inflicting suffering on most people who cannot afford to buy their way out, will require more than pledges. Shifting all transportation to zero-emission vehicles fueled by renewable electricity, ending the leakage of methane gas from agriculture, landfills, and gas fields, powering the economy with solar, wind, and geothermal energy all require investment. Charitable donations and voluntary action can lead the way, but only government regulation can overcome the fear of risk that is keeping most of the capital needed to make the transition quickly and with minimal disruption.

The good news is that the benefits of switching to a low-carbon economy go well beyond avoiding the worst effects of global climate change. As the contributors to this collection of essays show, there are a range of tangible improvements to the quality of life that come from the solutions they describe: healthier cities, accessible open space, more secure water supplies, good jobs. Not to mention the joy of reconnecting with nature, the source of our sense of well-being.

Acknowledgments

This book came into being thanks to Bill McKibben, Professor Mark Jacobson, Professor Robert Howarth, Dr. Nancy Ryan, Mary D. Nichols, Professor Jeff Dahn, Dr. Audrey Lee, Dr. David Barnes, Kate Gaertner, Laura Fedoruk, Michael Barnard, Dr. Luxmy Begum, and Steve Wheat. Thanks also to Anne Devlin, Deni Remsberg, Adrian Nagy, Elizabeth Nagy Loan, Ann Marie Hill, Cate Brown, Amalia Deligiannis, Kelly Falloon, Gaby Kalapos, Desislava Stefanova, Michelle Malcolm-Francis, William Pollock, and hundreds of climate-solution heroes who have gifted me with their stories.

Preface

Lakes, Trees, and Filthy Feet

By supper time on the last day of school each year, I'd be peeling off my shoes and socks before my dad's Volkswagen reached the muddy grass driveway at our summer cottage. The car windows would be rolled down, and my siblings would be hanging out of them, singing songs started by my mother or sister, happily drawing deep breaths into our lungs between each chorus.

The crisp spring air would already smell like alfalfa as we strained to catch the first glimpse of the blue-black water shimmering in the late day sun, with grey and pink seagulls dive-bombing for dinner. We'd be talking about chopping wood and having a bonfire and going fishing and leaping headfirst off the dock into the crystal blue lake, and that became, in hindsight, somehow more incredible than it was in the magic moment of the first swim. Racing down the wet beach sand, stumbling in blind excitement, we would fling ourselves into the waves, opening our eyes under the surface just in time to see the minnows and sunfish dart away in every direction.

After that, there would be about seventy glorious summer days in which we would do something, then go swimming, then do something else, then go swimming . . . It just never seemed to get old. I generally did not put my shoes back on until September, when we returned to the city for the new school year. The rejection of those shoes in June marked the beginning of the best dates on the annual calendar. They made me, made all of us, healthy, hardy, and hopeful, created a kind of internal fortitude so strong that everything the big bad world would throw at us over the decades to follow would just bounce off, like acorns bouncing off the branches and landing in the leaves on the woodpile.

Within a few days, my feet, and those of my four brothers, would sport some new colors. On top would be splotches of sunburn, but there was barely time to wince at the pain, let alone bother with the suntan cream my mother would try to have us apply. A few more days and the tops of our feet would turn satiny golden brown, while our toes and bottoms became Picasso paintings of calluses, cuts, bruises, and a deeply penetrated, ground-in ring of what my sister called "disgusting filth," although hers were dirty too.

What do you expect when you run through muck, sand, and gravel all day? No amount of swimming or scrubbing with a soapy brush would get rid of the dingy complexion from the bottoms of our feet. I mean, that's as far as I know, because, of course, there never seemed to be enough time for experimentation with cleanliness. Our feet would return to normal somewhere around Thanksgiving.

In the evenings, as the sun set below the trees on the perimeter of the yellow mustard growing row on row in the farm field behind our cottages, my sister Lizard (Elizabeth), would tease out the thorns and splinters with tweezers. We would ungratefully whine that she wasn't being gentle enough, while toasting our wieners and marshmallows over the fire, and asking my brother Dave to play a Simon & Garfunkel song on his twelve-string guitar.

Each purple abrasion and crimson contusion came with a story: how my toes were in the wrong place when the canoe decided to crash into the raft and squish two or three of them. Or how I met a big ugly gopher while running across a field and veered off the path into a thicket of thorns that we called prickles. Or how I cluelessly stepped on broken glass at the dump. When I would tell Lizard about the latter, I would whisper or just mouth the word "dump," because our parents said we weren't allowed to go there, which just guaranteed that we would go there.

Some time back then, New York folk troubadour Harry Chapin wrote an album called *Barefoot Boy*, and it became a favorite because I know in my heart I will always be a barefoot boy, thanks to two parents who had the good sense to invest in a modest scrap of land near Musselman's Lake with some crumbling shacks on it. They insisted that we would spend our summers there. Our flimsy initial objections were quickly lost somewhere in a montage of swimming, fishing, waterskiing, tree forts, pony rides, apple picking, corn roasts, Saturday night dances in the bingo hall, or over at the commercial beach, loud live rock music, romantic jukebox ballads, and first loves.

It was a tight little community where everyone yelled at us kids to behave, but they also kept us safe and helped us create the memories that we didn't know we were making, didn't realize would matter to us in later years. It's a little like the history we're rewriting now with climate solutions.

We, my family, and indeed, we the human species, have always craved nature. Even as we have built things. Cities and industry may be monuments to our ingenuity, creativity, and brainpower, but it is to the leafy forests, sparkling lakes, and starry summer skies that we retreat to soothe our souls. It is in nature that we feel peace and well-being; in places where we can breathe the kind of air only trees make, taste water from inside the earth, smell wildflowers, and hear the buzzing of heat bugs and bees rather than power tools. Our health and happiness are found in places where there are fewer human-made things, not more. It's an assault on our core humanity to wipe these places out. None of us, not even the profit-takers, prefers air fouled by gaseous fumes over the fresh breezes near the rivers and woodlands.

We must save this natural world, that we spare so little time for, instead of rushing toward goals which may seem less important as we grow older and realize our loves, friends, families, and green spaces should have mattered more. The earth was one such green place before humans took control, and it will be again after we're gone. Could it be that the only choice we have left is how soon we want our own extinction to happen?

Right now, we're hurtling at breakneck speed toward trouble, like an inebriated twenty-year-old in his dad's pickup truck. Sadly, there might be an innocent, wide-eyed baby sister in the back seat. Why not pull over up here by the lake? Step out of the truck, stand by the water, and take a few deep breaths. Lift our baby sister out of the back, hold her hand, and show her the stars. Maybe even take off our shoes and feel the wet grass between our toes. Because, in the end, we're all barefoot boys and girls. That's how we were born.

Introduction

The Need for Speed

We need answers, and we need speed. Can we electrify all our buildings and transportation without crushing the grid or breaking the bank? Has anyone yet built a town that can survive huge storms and wildfires? Do heat pumps actually work? How green are electric vehicles? Do we need carbon capture, nuclear power, and hydrogen? How is sewer water reducing emissions? What is a virtual power plant, and why does it matter? How will the circular economy function, specifically? This book answers these questions and dozens more.

* * *

If you were anywhere in the universe, on any night, you could probably look up and see orbital bodies in the sky: twinkling and shooting stars against the deep midnight-blue, refracted colors, or hazy clouds. Although you might see hundreds of stars from wherever you were, we're told no other planets in the Milky Way galaxy would support human and other life quite like ours does. That's astonishing because there are billions of planets here. The galaxy is about 220,000 light years wide. Just one light year is 5.9 trillion miles long.

The experts say that if there is another planet that could work, we can't move our populations there easily with today's space technology, or if we can, we might be embarking on an untenable experience, such as crowding into a dome on the moon or Mars. To go safely into the galaxy beyond, we need far faster spaceships. Such a scenario is filled with unknowns and wouldn't become real with the snap of our fingers. It's too soon to say otherwise—this is our best bet for a viable home. We're children of a unique and special place called Earth.

That's reality, and so is this: Our planet is on fire. The fires are multiplying, getting bigger, hotter, more devastating. The heat domes, too. The storms and floods grow worse. These are facts. Most recently, the negative upward trend-lines have literally been going off the charts, far beyond projections. Anyone with eyes can see this crisis is not minor, temporary, or far off in the future. It will soon affect all our families. If we don't act rapidly, we'll be killed, sickened, or injured, along with pets, other animals, trees, plants, water, homes.

THE CLIMATE CRISIS

The climate crisis imperative is to halt this downward spiral now, while we still can. The UN Secretary-General, António Guterres, former US Vice President Al Gore, and Greta Thunberg are right. This should be called a "crisis" or "emergency." Those who don't agree are the "radicals" among us. The imperative is for bold, full-frontal government action, prioritizing rapid deployment of proven solutions. It appears we will have little choice but to remove many governments because they are now dangerous.

> "The climate crisis imperative is to halt this downward spiral now, while we still can."

In technical language, indicative projections have surged beyond history's most significant extinction event, the Permian–Triassic, which wiped out 70 percent of terrestrial vertebrates, 81 percent of marine species, 83 percent of genera, and 57 percent of biological families. We're heading towards extinction or near extinction. Earth's temperatures were fairly stable for about seven thousand years. The last one hundred years have been different. Temperatures are increasing so fast that scientists are panicking. Every year is hotter, and records are breaking by bigger margins, more frequently.

Wildfires are erasing millions of acres of forests and cities. Millions will die, are dying from heat, air pollution, and new pandemics. We're losing the forests, agriculture, food and water, the river-feeding glaciers, the North and South Poles, planetary stability, ocean currents, everything that gave us safety, health, and quality of life. Millions are losing homes to unlivable heat, storms, and flooding. Some countries are trying to prevent migrations across their borders. The heat, fires, storms, disease, and water shortages will lead to violence and death. These are facts.

We need solutions that can make big impacts now, not more talk about untested ideas. We need the clear information in this book to defog our brains and focus our efforts on productive forward momentum along specific pathways. We need finalized priorities and swift action by 2035. This book focusses on 2024–2035. And government and media people should read this. The world needs your leadership. There's no time left to get it wrong.

Right now, we need good judgement and decisiveness. No more technobabble about compromises that might help. Let's focus on high-impact, proven solutions. No more amplification of opinions from the wrong people. Who is bankrolling this person? Why are they saying this? Who are the sincere, qualified experts that know what's proven and which are theoretical someday schemes dreamed up to land a slush pile of government research dollars, negotiate regulations, and confuse the public?

> "We need the clear information in this book to defog our brains and focus. And government and media people should read this."

PROVEN SOLUTIONS

Proven solutions are based on evidence and experience. They make an impact, not just once but sustainably. They're socio-economically feasible, technically viable, and scalable by 2035. Those who say, "We must do all of the above" are not experts, have not done their homework, don't know which technologies will make an impact fast. They're compromised, afraid, incrementalists. Delay is now the key challenge.

And make no mistake, at root, the climate crisis is a technology problem. It was created by outdated technology, and it will be solved with modern technology, plus programs, economic strategies, and politics that enable implementation of modern technology. Those who say otherwise don't really understand the polluting of air, water, and land. At its core, it's a technology problem.

> "What's a proven solution? It's socio-economically feasible, technically viable, and scalable by 2035."

Should we "pick winner and loser" technologies? Absolutely, yes. If it had not been delayed by profiteers, engineers would have corrected this technology problem by now. In Norway, citizens still trust their experts rather than compromised politicians. Despite the country being a big oil producer, about 80 percent of the people now buy electric vehicles and heat pumps. Norwegians don't let fake news from corrupt people fog their brains. When a politician says he or she doesn't want to "pick winners," it means they don't want to anger their political donors.

We're battling an unpredictable, dangerous enemy that is not only outside our gates but its traitors, villains, and profit-takers are also moving among us. They're inside our homes, chatting up our children, and they're inside our heads. The expert plans in these pages can help with the problem inside our heads.

EXPERTS

Just as we have no other home in the vast Milky Way, there is also no one else like Bill McKibben in our galaxy. He has written twelve books on environmental economics. Professor Mark Jacobson of Stanford is also one of a kind. He counsels the White House (three administrations) and has created clean energy strategies for 145 countries and all fifty states. Professor Nancy Ryan was the California Public Utilities Commissioner. Few can offer a more realistic understanding of how we will proceed with electrifying our cars, trucks, and other transportation systems and seamlessly power them with a modernized grid. Only Professor Robert Howarth at Cornell has so definitively exposed the scientific truth about fracking first and, later, blue hydrogen.

You won't find many people who know more about generating solar energy and optimizing solar resources than Dr. Audrey Lee from Sunrun, or anyone who can write green business roadmaps like Kate Gaertner for local governments, food systems, and large corporations such as Nike and StockX. Professor Jeff Dahn's team at Dalhousie works for Tesla, developing technology that's now being adopted by the major electric vehicle and battery makers. Dr. David Barnes, winner of the 2023 Polar Medal, sent an article for this book from a research vessel in Antarctica. He says improvements are needed to 30 by 30, the UN plan for protection of 30 percent of land and sea by 2030.

These people are not guessing or musing. They're among the top people in the field, and they do this full-time. They know for certain what works and what needs to be done. If you search online for "climate solutions," you'll find a light-years long line of individuals who think they're experts, but they may be quite unlike the people whose words are found here. These writers have spent their entire careers learning about exactly how we will save humanity and the natural environment from destructive forces.

> "These people are not guessing or musing. They're among the top people in the field."

Our thinking problem has meant that we don't have a consensus and a clear, sustained, unified message for our leaders. We've become conditioned to cope by regularly wiping out our memory banks. We seem to be living in the movie *Don't Look Up*. We spend about three seconds assessing each of the thousands of ideas fighting for our attention every day, as if we're strolling through the central marketplace, with merchants offering a taste of the weekly cheese special or showing off a great new gadget.

For most of us, this marketplace is online, its merchants cheery little bings, notifications we reflexively tap. By the time we're halfway through our morning coffees, we've scanned six items, each asking more questions than they answer, with intelligent-sounding insights in the comments section. Where did all these experts come from? They pluck an interesting fact or two from Wikipedia or Google and plug them into a funny rhetorical remark, then move on to the next unsuspecting victim.

A reasoned exchange of ideas has not taken place. It's like thinking you've been destroyed by some schoolyard comment that made the cool kids laugh. But you have not been destroyed, and if you keep thinking that, sooner or later a big sister, parent, or professional psychologist will tell you, "The problem is inside your head." Uncle Joe, who says gas cars are greener than electric cars, is not an expert. The gas furnace guy who says heat pumps don't work just likes beer more than training.

True reason involves research, evidence, and peer review. An expert is someone who has studied the field for years, not for minutes. They've conducted hundreds of experiments and read thousands of reports. They're familiar with approaches, alternative strategies, salient facts, and the weight of evidence. And every day they learn more about their areas of study. They've cringed at all the ridiculous opinions from online outdated pseudo experts. They may employ rhetoric to keep things lively, but they usually add a footnote so that you know they can back up what they're saying.

They work for universities and media firms, not for oil, gas, chemical, or nuclear companies. And they're not corporate-funded politicians. They reach their conclusions on the basis of rigorous investigations, experimentation, and the scientific method. Half the writers in this book are scientists who publish in peer-reviewed journals. Their paychecks remain the same no matter what they discover and report, and they are happy with that. The other contributors are professional writers who will likely be fired if even one of their stories leads to a lawsuit over inaccurate facts. The planet is in crisis, and so are we. It's time to listen to well-intentioned experts.

The history of human survival on Earth is nearly impossible. At some point, we were few in number and up against big, fast, vicious land and sea creatures with sharp teeth and claws who wanted to eat us for a midnight snack while we and our children lay sleeping in a forest or cave. Somehow, we outran them. We summoned the adrenaline, courage, and physical stamina to fight them off and flee to higher ground where we could rest and think about how to avoid death for a few more days. We began with instincts for fight or flight, lived in the moment, became strong, quick, and clever. When it was us against the animals, speed, adrenaline, and a bit of intelligence was enough to keep going.

As we developed other strengths like imagination, ingenuity, and teamwork, we dominated and multiplied, and we thought we were pretty fantastic. We never lost our need for speed, our wanderlust, or our resolute drive to survive, even as we became educated and sophisticated. Did we learn to move beyond the moment and properly channel our energy? Some of us, perhaps.

Our drive to thrive became a mess. Adrenaline became our nemesis. Our relentless obsession with a better life, to have more, have it now, cure it now, eat it now; our craving for acceleration and convenience has warped our brains or made them narrow. We're on autodrive. We don't know where we're going. We're just going. And the Earth is too small for our numbers and our destructive tendencies. The fires are closing in. We're running out of safe places and running out of time. Let's turn destructive speed into a fuel for big, fast fixes.

We need innovation, cooperation, and efficiency, a consensus on our plan, communicated loudly and clearly. Let's swiftly eliminate the opposing forces that are driven by blinding self-interest. They can't seem to see beyond the moment.

The climate menace is formidable. It's as overwhelming as a million-acre forest fire, as brutal as a five-year drought, as furious as a forty-foot ocean tsunami. This is not a Hollywood movie. Your SUV and firearms will not save your family. A SWAT team is not coming to rescue you. When your life is taken during crisis, you don't usually see it coming, like the souls trapped in their cars when Perris, California, was engulfed in flames. Danger might be closer at hand than you realize. Knowing what the right solutions are and deploying them quickly is now really a matter of life and death: my life, your life, and the lives of our family members.

> "We have never lost our need for speed, our wanderlust, or our resolute drive to survive. Let's properly channel this energy."

Chapter One

Stop Burning Things

The Truth Is New and Counterintuitive: We Have the Technology Necessary to Rapidly Ditch Fossil Fuels

Bill McKibben

William Ernest (Bill) McKibben is an American author, journalist, and environmentalist who has written extensively on global warming and climate change. As detailed in his biography near the end of this book, he has written more than a dozen books and hundreds of feature magazine articles about the environment. He has organized several movements, including campaigns which motivated top-level financial people to make high-impact decisions affecting energy investment; and also 350.org's grass-roots efforts, which resulted in 5,200 simultaneous demonstrations in 181 countries in 2009 and at least 7,000 events in 188 countries in 2010. In the 2020s, he is mobilizing people worldwide through the Third Act campaign.

We don't know when or where humans started building fires; as with all things primordial, there are disputes. But there is no question of the moment's significance. Fire let us cook food, and cooked food delivers far more energy than raw; our brains grew even as our guts, with less processing work to do, shrank. Fire kept us warm, and human enterprise expanded to regions that were otherwise too cold. And, as we gathered around fires, we bonded in ways that set us on the path to forming societies. No wonder Darwin wrote that fire was "the greatest discovery ever made by man, excepting language."

> "Having spent millennia learning to harness fire, and three centuries using it to fashion the world we know, we must spend the next years systematically eradicating it."

Darwin was writing in the years following the Industrial Revolution, when we learned how to turn coal into steam power, gas into light, and oil into locomotion, all by way of combustion. Our species depends on combustion; it made us human, and then it made us modern. But, having spent millennia learning to harness fire, and three centuries using it to fashion the world we know, we must spend the next years systematically eradicating it. Because, taken together, those blazes—the fires beneath the hoods of 1.4 billion vehicles and in the homes of billions more people, in giant power plants, and in the boilers of factories and the engines of airplanes, ships—are more destructive than the most powerful volcanoes, dwarfing Krakatoa and Tambora.

The smoke and smog from those engines and appliances directly kill nine million people a year, more deaths than those caused by war and terrorism, not to mention malaria and tuberculosis, together. (In 2020, fossil-fuel pollution killed three times as many people as COVID-19 did.) Those flames, of course, also spew invisible and odorless carbon dioxide at an unprecedented rate; that CO_2 is already rearranging the planet's climate, threatening not only those of us who live on it now but all those who will come after us.

But here's the good news, which makes this exercise more than merely rhetorical: rapid advances in clean-energy technology mean that all that destruction is no longer necessary. In the place of those fires we keep lit day and night, it's possible for us to rely on the fact that there is a fire in the sky—a great ball of burning gas about ninety-three million miles away, whose energy can be collected in photovoltaic panels and which differentially heats the Earth, driving winds whose energy can now be harnessed with great efficiency by turbines. The electricity they produce can warm and cool our homes, cook our food, and power our cars and bikes and buses. The sun burns so we don't need to.

Wind and solar power are not a replacement for everything, at least not yet. Three billion people still cook over fire daily and will at least continue to do so until sufficient electricity reaches them, and perhaps thereafter, since culture shifts slowly. Even then, flames will still burn—for birthday-cake candles, for barbecues, for joints (until you've figured out the dosing for edibles)—just as we still use bronze, though its age has long passed. And there are a few larger industries—intercontinental air travel, certain kinds of metallurgy such as steel production, and so on—that may require combustion, probably of hydrogen, for some time longer. But these are relatively small parts of the energy picture. And in time, they, too, will likely be replaced by renewable electricity. (Electric-arc furnaces are already producing some kinds of steel, and Japanese researchers have just announced a battery so light that it might someday power passenger flights across oceans.) In fact, I can see only one sublime, long-term use for large-scale planned combustion, which I will get to. Mostly, our job as a species is clear: stop smoking.

As of 2022, this task is both possible and affordable. We have the technology necessary to move fast, and deploying it will save us money. Those are the first key ideas to internalize. They are new and counterintuitive, but a few people have been working to realize them for years, and their stories make clear the power of this moment.

When Mark Jacobson was growing up in northern California in the 1970s, he showed a gift for science and also for tennis. He travelled for tournaments to Los Angeles and San Diego, where, he told me recently, he was shocked by how dirty the air was: "You'd get scratchy eyes, your throat would start hurting. You couldn't see very far. I thought, *Why should people live like this?*" He eventually wound up at Stanford, first as an undergraduate and then, in the mid-1990s, as a professor of civil and environmental engineering, by which time it was clear that visible air pollution was only part of the problem. It was understood that the unseen gas produced by combustion—carbon dioxide—posed an even more comprehensive threat.

> "This task is both possible and affordable. We have the technology necessary to move fast, and deploying it will save us money."

To get at both problems, Jacobson analyzed data to see if an early-model wind turbine sold by General Electric could compete with coal. He worked out its capacity by calculating its efficiency at average wind speeds; a paper he wrote, published in the journal *Science* in 2001, showed that you "could get rid of sixty per cent of coal in the U.S. with a modest number of turbines." It was, he said, "the shortest paper I've ever written—three-quarters of a page in the journal—and it got the most feedback, almost all from haters." He ignored them; soon, he had a graduate student mapping wind speeds around the world, and then he expanded his work to other sources of renewable energy.

In 2009, he and Mark Delucchi, a research scientist at the University of California, published a paper suggesting that hydroelectric, wind, and solar energy could conceivably supply enough power to meet all the world's energy needs. The conventional wisdom at the time was that renewables were unreliable because the sun insists on setting each night and the wind can turn fickle. In 2015, Jacobson wrote a paper for the *Proceedings of the National Academy of Sciences*, showing that, on the contrary, wind and solar energy could keep the electric grid running.

Time is proving Jacobson correct: a few nations—including Iceland, Costa Rica, Namibia, and Norway—are already producing more than 90 percent of their electricity from clean sources. When Jacobson began his work, wind turbines were small fans atop California ridgelines, whirligigs that looked more like toys than power sources. Now General Electric routinely erects windmills about three times as tall as the Statue of Liberty, and in August 2023, a Chinese firm announced a new model, whose blades will sweep an area the size of six soccer fields, with each turbine generating enough power for twenty thousand homes. (An added benefit: bigger turbines kill fewer birds than smaller ones, though, in any event, tall buildings, power lines, and cats are responsible for far more avian deaths.) In December, Jacobson's Stanford team published an updated analysis, stating that we have 95 percent of the technology required to produce 100 percent of America's power needs from renewable energy by 2035, while keeping the electric grid secure and reliable.

> **"We have 95 percent of the technology required to produce 100 percent of America's power needs from renewable energy by 2035."**

Making clean technology affordable is the other half of the challenge, and here the news is similarly upbeat. In September, after almost fifteen years of work, a team of researchers at Oxford University released a paper that is currently under peer review but which, fifty years from now, people may look back on as a landmark step in addressing the climate crisis. The lead author of the report is Oxford's Rupert Way; the research team was led by an American named Doyne (pronounced "dough-en") Farmer.

Farmer grew up in New Mexico, a precocious physicist and mathematician. His first venture, formed while he was a graduate student at UC Santa Cruz, was called Eudaemonic Enterprises, after Aristotle's term for the condition of human flourishing. The goal was to beat roulette wheels. Farmer wore a shoe (now housed in a German museum) with a computer in its sole and watched as a croupier tossed a ball into a wheel; noting the ball's initial position and velocity, he tapped his toe to send the information to the computer, which performed quick calculations, giving him a chance to make a considered bet in the few seconds the casino allowed.

This achievement led him to building algorithms to beat the stock market—a statistical-arbitrage technique that underpinned an enterprise he co-founded called the Prediction Company, which was eventually sold to the Swiss banking giant UBS. Happily, Farmer eventually turned his talents to something of greater social worth: developing a way to forecast rates of technological progress.

The basis for this work was research published in 1936, when Theodore Wright, an executive at the Curtiss Aeroplane Company, had noted that every time the production of airplanes doubled, the cost of building them fell by 20 percent. Farmer and his colleagues were intrigued by this "learning curve" (and its semiconductor-era variant, Moore's Law); if you could figure out which technologies fit on the curve and which didn't, you'd be able to forecast the future.

"It was about fifteen years ago," Farmer told me, in December 2021. "I was at the Santa Fe Institute, and the head of the National Renewable Energy Lab came down. He said, 'You guys are complex-systems people. Help us think outside the box—what are we missing?' I had a Transylvanian postdoctoral fellow at the time, and he started putting together a database—he had high-school kids working on it, kids from St. John's College in Santa Fe, anyone. And, as we looked at it, we saw this point about the improvement trends being persistent over time."

The first practical application of solar electricity was on the Vanguard 1 satellite, in 1958—practical if you had the budget of the space program. Yet the cost had been falling steadily, as people improved each generation of the technology—not because of one particular breakthrough or a single visionary entrepreneur but because of constant incremental improvement. Every time the number of solar panels manufactured doubles, the price drops another 30 percent, which means that it's currently falling about 10 percent every year.

But—and here's the key—not all technologies follow this curve. "We looked at the price of coal over 140 years," Farmer said. "Mines are much more sophisticated, the technology for locating new deposits is much better. But prices have not come down." A likely explanation is that we got to all the easy stuff first: oil once bubbled up out of the ground; now we have to drill deep beneath the ocean for it. Whatever the reason, by 2013, the cost of a kilowatt-hour of solar energy had fallen by more than 99 percent since it was first used on the Vanguard 1. Meanwhile, the price of coal has remained about the same. It was cheap to start, but it hasn't gotten cheaper.

The more data sets that Farmer's team members included, the more robust numbers they got, and by the autumn of 2021, they were ready to publish their findings. They found that the price trajectories of fossil fuels and renewables are already crossing. Renewable energy is now cheaper than fossil fuel and becoming more so. So a "decisive transition" to renewable energy, they reported, would save the world twenty-six trillion dollars in energy costs in the coming decades.

The constant price drops mean, Farmer said, that we might still be able to move quickly enough to meet the target set in the 2016 Paris Climate Accords of trying to limit temperature rise to 1.5 degrees Celsius. "One point five is going to suck," he said. "But it sure beats three. We just need to put our money down and do it. So many people are pessimistic and despairing, and we need to turn that around."

Numbers like Farmer's make people who've been working in this field for years absolutely giddy. At the 2021 United Nations Climate Change Conference (COP26), I retreated one day from Glasgow's giant convention center to the relative quiet of the city's university district for a pizza with a man named Kingsmill Bond. Bond is an Englishman and a former investment professional, and he looks the part: lean, in a bespoke suit, with a good haircut. His daughter, he said, was that day sitting her exams for Cambridge, the university he'd attended before a career at Citi and Deutsche Bank that had

Bill McKibben

taken him to Hong Kong and Moscow. He'd quit some years ago, taking a cut in pay that he's too modest to disclose. He'd worked first for the Carbon Tracker Initiative, in London, and now the Rocky Mountain Institute, based in Colorado, two groups working on energy transition.

He drew on a napkin excitedly, expounding on the numbers in the Oxford report. We would have to build out the electric grid to carry all the new power, and install millions of electric vehicle (EV) chargers, and so on, down a long list—amounting to maybe a trillion dollars in extra capital expenditure a year over the next two or three decades. But, in return, Bond said, we get an economic gift: "We save about two trillion dollars a year on fossil-fuel rents. Forever."

> "It would be far less expensive than not moving fast—that's measured in hundreds of trillions of dollars but also in millions of lives."

Even if you're pessimistic about how much it will cost to make the change, though, it's clear that it would be far less expensive than not moving fast—that's measured in hundreds of trillions of dollars but also in millions of lives and whatever value we place on maintaining an orderly civilization.

The new numbers turn the economic logic we're used to upside down. A few years ago, at a petroleum-industry conference in Texas, the Canadian Prime Minister, Justin Trudeau, said something both terrible and true: that "no country would find 173 billion barrels of oil in the ground and leave them there." He was referring to Alberta's tar sands, where a third of Canada's natural gas is used to heat the oil trapped in the soil sufficiently to get it to flow to the surface and separate it from the sand. Just extracting the oil would put Canada over its share of the carbon budget set in Paris, and actually burning it would heat the planet nearly half a degree Celsius and use up about a third of the total remaining budget. (And Canadians account for only about one half of 1 percent of the world's population.)

Even on purely economic terms, such logic makes less sense with each passing quarter. That's especially true for the 80 percent of people in the world who live in countries that must import fossil fuels—for them, it's all cost and no gain. Even for petrostates, however, the spreadsheet is increasingly difficult to rationalize. Bond supplied some numbers: Canada has fossil-fuel reserves totalling 167 petawatt hours, which is a lot. (A petawatt is a quadrillion watts.) But, he said, it has potential renewable energy from wind and solar power alone of seventy-one petawatt hours a year. A reasonable question to ask Trudeau would be: What kind of country finds a windfall like that and simply leaves it in the sky?

It's worth remembering that a transition to renewable energy would, by some estimates, reduce the total global mining burden by as much as 80 percent, because so much of what we dig up today is burned (and then we have to go dig up some more). You dig up lithium once and put it to use for decades in a solar panel or battery. In fact, a switch to renewable energy will reduce the load on all kinds of systems.

At the moment, roughly 40 percent of the cargo carried by ocean-going ships is coal, gas, oil, and wood pellets—a never-ending stream of vessels crammed full of stuff to burn. You need a ship to carry a wind turbine blade, too, if it's coming from across the sea, but you only need it once. A solar panel or a windmill, once erected, stands for a quarter of a century or longer. The US military is the world's largest single consumer of fossil fuels, but 70 percent of its logistical "lift capacity" is devoted solely to transporting the fossil fuels used to keep the military machine running.

> **"Roughly 40 percent of the cargo carried by ocean-going ships is coal, gas, oil, and wood pellets—a never-ending stream of vessels crammed full of stuff to burn."**

Raw materials aren't the only possible pinch point. We're also short of some kinds of expertise. Saul Griffith is perhaps the world's leading apostle of electrification. (His 2021 book is titled *Electrify*.) An Australian by birth, he has spent recent years in Silicon Valley, rallying entrepreneurs to the project of installing EV chargers, air-source heat pumps, induction cooktops, and the like. He can show that they save homeowners, landlords, and businesses money; he's also worked out the numbers to show that banks can prosper by extending, in essence, mortgages for these improvements. But he told me that, to stay within the 1.5-degree Celsius range, "America is going to need a million more electricians this decade." That's not impossible. Working as an electrician is a good job, and community colleges and apprenticeship programs could train many more people to become one. But, as with the rest of the transition, it's going to take leadership and coordination to make it happen.

Change on this scale would be difficult even if everyone was working in good faith, and not everyone is. So far, for instance, the climate provisions of the Build Back Better Act, which would help provide, among many other things, training for renewable-energy installers, have been blocked not just by the oil-dominated GOP but also by Joe Manchin, the Democrat who received more fossil-fuel donations in the past election cycle than anyone else in the Senate. The thirty-year history of the global-warming fight is largely a story of the efforts by the fossil-fuel industry to deny the need for change, or, more recently, to insist that it must come slowly.

The fossil-fuel industry wants to be able to keep burning something. That way, it can keep both its infrastructure and its business model usefully employed. It's like an industry of rational pyromania. A decade or so ago, the thing it wanted to burn next was natural gas. Since it produces less carbon dioxide than coal does, it was billed as the "bridge fuel" that would get us to renewables. The logic seemed sound. But researchers, led by Bob Howarth, at Cornell University, found that producing large quantities of natural gas released large quantities of methane into the atmosphere. And methane (CH_4) is, like CO_2, a potent heat-trapping gas, so it's become clear that natural gas is a bridge fuel to nowhere—clear, that is, to everyone but the industry. The head of a big gas firm told a conference in Texas in 2022 that he thought the domestic gas industry could be producing for the next hundred years.

Other parts of the industry want to go further back in time and burn wood; the European Union and the United States officially class "biomass burning" as carbon neutral. The city of Burlington, in my home state of Vermont, claims to source all its energy from renewables, but much of its electricity comes from a plant that burns trees. Again, the logic originally seemed sound: If you cut a tree, another grows in its place, and it will eventually soak up the carbon dioxide emitted from burning the first tree. But, again, "eventually" is the problem. Burning wood is highly inefficient, and so it releases a huge pulse of carbon right now, when the world's climate system is most vulnerable.

Trees that grow back in a few generations' time will come too late to save the ice caps. The world's largest wood-burning plant is in England, run by a company called Drax; the plant used to burn coal, and it does scarcely less damage now than it did then. In January 2022, news came that Enviva, a company based in Maryland that is the largest producer of wood pellets in the world, plans to double its output.

> "Trees that grow back in a few generations' time will come too late to save the ice caps."

Or consider the huge sums of money in the bipartisan infrastructure bill passed in 2022, which will support another technology called carbon capture. This involves fitting power plants with enough filters and pipes so that they can go on burning coal or gas but capture the CO_2 that pours out of the smokestacks and pipe it safely away—into an old salt mine, perhaps. (Or, ironically, into a depleted oil well, where it may be used to push more crude to the surface.) So far, these carbon-capture schemes don't really work—but, even if they did, why spend the money to outfit systems with pipes and filters when solar power is already cheaper than coal power? We will have to remove some of the carbon in the atmosphere, and new generations of direct-air-capture machines may someday play a role, if their cost drops quickly. (They use chemicals to filter carbon straight from the ambient air; think of them as artificial trees.) But using this technology to lengthen the lifespan of coal-fired power plants is just one more gift to a politically connected industry.

Increasingly, the fossil-fuel industry is turning toward hydrogen as an out. Hydrogen does burn cleanly without contributing to global warming, but the industry likes hydrogen because one way to produce it is by burning natural gas. And, as Howarth and Jacobson demonstrated in a recent paper, even if you combine burning that gas with expensive carbon capture, the methane that leaks from the frack wells is enough to render the whole process ruinous environmentally, and it makes no sense economically without huge subsidies.

There is another way to produce hydrogen, and, in time, it will almost certainly fuel the last big artificial fires on our planet. Through electrolysis, hydrogen can be separated from oxygen in water. And if the electricity used in the process is renewably produced, then this "green hydrogen" would allow countries such as Japan, Singapore, and Korea, which may struggle to find enough space in their landscapes for renewable-energy generation, to power their grids. The Australian billionaire Andrew Forrest, the founder of the Fortescue Metals Group, is proposing to use solar power to produce green hydrogen that he can then ship to those countries. In January, Mukesh Ambani, the head of Reliance Industries and the richest man in India, announced plans to spend seventy-five billion dollars on the technology. Airbus recently predicted that green hydrogen could fuel its long-haul planes by 2035. And the good news—though Farmer cautions that the data sets are still pretty scanty—is that the electrolysers which use solar energy to produce hydrogen seem to be on the same downward cost curve as solar panels, wind turbines, and batteries.

The fossil-fuel industry can be relied on to fight these shifts. Last autumn, a utility company in Oklahoma announced that it would charge fourteen hundred dollars to disconnect residential gas lines and move home stoves and furnaces to electricity. Within days, other utilities followed suit. That's why the climate movement is increasingly taking on the banks that make loans for the expansion of fossil-fuel infrastructure.

Last year, the International Energy Agency said that such expansion needed to end immediately if we are to meet the Paris targets, yet the world's biggest banks, while making noises about "net zero by 2050," continue to lend to new pipelines and wells.

The shift away from combustion is large and novel enough that it bumps up against everyone's prior assumptions—environmentalists, too. The fight against nuclear power, for example, was an early mainstay of the green movement, because it was easy to see that if something went wrong it could go badly wrong. I applauded, more than a decade ago, when the Vermont legislature voted to close the state's old nuclear plant at the end of its working life, but I wouldn't today. Indeed, for some years, I've argued that existing nuclear reactors that can still be run with any margin of safety probably should be, as we're making the transition—the spent fuel they produce is an evil inheritance for our descendants, but it's not as dangerous as an overheated Earth, even if the scenes of Russian troops shelling nuclear plants added to the sense of horror enveloping the planet these past weeks. Yet the rapidly falling cost of renewables also indicates why new nuclear plants will have a hard time finding backers; it's evaporating nuclear power's one big advantage—that it's always on. Farmer's Oxford team ran the numbers. "If the cost of coal is flat, and the cost of solar is plummeting, nuclear is the rare technology whose cost is going up," he said. Advocates will argue that this is because safety fears have driven up the cost of construction. "But the only place on Earth where you can find the cost of nuclear coming down is Korea," Farmer said. "Even there, the rate of decline is 1 percent a year. Compared to 10 percent for renewables, that's not enough to matter."

Accepting nuclear power for a while longer is not the only place environmentalists will need to bend. A reason I supported shutting down Vermont's nuclear plant was because campaigners had promised that its output would be replaced with renewable energy. In the years that followed, though, advocates of scenery, wildlife, and forests managed to put the state's mountaintops off-limits to wind turbines. More recently, the state's public-utility commission blocked construction of an eight-acre solar farm on aesthetic grounds. Those of us who live in and love rural areas have to accept that some of that landscape will be needed to produce energy. Not all of it, or even most of it—Jacobson's latest numbers show that renewable power actually uses less land than fossil fuels, which require drilling fifty thousand new holes every year in North America alone. But we do need to see our landscape differently—as Ezra Klein wrote in March 2022 in the *New York Times*, "to conserve anything close to the climate we've had, we need to build as we've never built before."

Ethanol: "You could produce eighty times the amount of automobile mileage using an equivalent area of land (for solar)."

Corn fields, for instance, are a classic American sight, but they're also just solar-energy collectors of another sort. (And ones requiring annual applications of nitrogen, which eventually washes into lakes and rivers, causing big algae blooms.) More than half the corn grown in Iowa actually ends up as ethanol in the tanks of cars and trucks—in other words, those fields are already growing fuel, just inefficiently. Because solar panels are far more efficient than photosynthesis, and because EVs are far more efficient than cars with gas engines, Jacobson's data show that, by switching from ethanol to solar, you could produce eighty times the amount of automobile mileage using an equivalent area of land. And the transition could bring some advantages: the market for electrons is predictable, so solar panels can provide a fairly stable income for farmers, some of whom are learning to grow shade-tolerant crops or to graze animals around and beneath them.

Whenever I write about the rise of EVs, Twitter responds that we'd be better off riding bikes and electric buses. In many ways, we would be, and some cities are thankfully starting to build extensive bike paths and rapid-transit lanes for electric buses. But, as of 2017, just 2 percent of passenger miles in this country come from public transportation. Bike commuting has doubled in the past two decades—to about 1 percent of the total. We could (and should) quintuple the number of people riding bikes and buses, and even then, we'd still need to replace tens of millions of cars with EVs to meet the targets in the time the scientists have set to meet them. That time is the crucial variable. As hard as it will be to rewire the planet's energy system by decade's end, I think it would be harder—impossible, in fact—to sufficiently rewire social expectations, consumer preferences, and settlement patterns in that short stretch.

So one way to look at the work that must be done with the tools we have at hand is as triage. If we do it quickly, we will open up more possibilities for the generations to come. Just one example: Farmer says that it's possible to see the cost of nuclear-fusion reactors, as opposed to the current fission reactors, starting to come steeply down the cost curve—and to imagine that within a generation or two people may be taking solar panels off farm fields because fusion (which is essentially the physics of the sun brought to Earth) may be providing all the power we need. If we make it through the bottleneck of the next decade, much may be possible.

There is one ethical element of the energy transition that we can't set aside: the climate crisis is deeply unfair—by and large, the less you did to cause it, the harder and faster it hits you—but in the course of trying to fix it we do have an opportunity to also remedy some of that unfairness. For Americans, the best part of the Build Back Better bill may be that it tries to target significant parts of its aid to communities hardest hit by poverty and environmental damage, a residue of the Green New Deal that is its parent. And advocates are already pressing to insure that at least some of the new technology is owned by local communities—by churches and local development agencies, not by the solar-era equivalents of Koch Industries or ExxonMobil.

Advocates are also calling for some of the first investments in green transformations to happen in public-housing projects, on reservations, and in public schools serving low-income students. There can be some impatience from environmentalists who worry that such considerations might slow down the transition. But, as Naomi Klein recently told me, "The hard truth is that environmentalists can't win the emission-reduction fight on our own. Winning will take sweeping alliances beyond the self-identified green bubble—with trade unions, housing-rights advocates, racial-justice organizers, teachers, transit workers, nurses, artists, and more. But, to build that kind of coalition, climate action needs to hold out the promise of making daily life better for the people who are most neglected right away—not far off in the future. Green, affordable homes and water that is safe to drink is something people will fight for a hell of a lot harder than carbon pricing."

These are principles that must apply around the world, for basic fairness and because solving the climate crisis in just the United States would be the most pyrrhic of victories. (They don't call it "global warming" for nothing.) In Glasgow, I sat down with Mohamed Nasheed, the former President of the Maldives and the current speaker of the People's Majlis, the nation's legislative body. He has been at the forefront of climate action for decades because the highest land in his country, an archipelago that stretches across the equator in the Indian Ocean, is just a few meters above sea level.

At COP26, he was representing the Climate Vulnerable Forum, a consortium of fifty-five of the nations with the most to lose as temperatures rise. As he noted, poor countries have gone deeply into debt trying to deal with the effects of climate change. If they need to move an airport or shore up seawalls, or recover from a devastating hurricane or record rainfall, borrowing may be their only recourse. And borrowing gets harder, in part, because the climate risks mean that lenders demand more. The climate premium on loans may approach 10 percent, Nasheed said; some nations are already spending 20 percent of their budgets just paying interest. He suggested that it might be time for a debt strike by poor nations.

The rapid fall in renewable-energy prices makes it more possible to imagine the rest of the world chipping in. So far, though, the rich countries haven't even come up with the climate funds they promised the Global South more than a decade ago, much less any compensation for the ongoing damage that they have done the most to cause. (All of sub-Saharan Africa is responsible for less than 2 percent of the carbon emissions currently heating the earth; the United States is responsible for 25 percent.)

Tom Athanasiou's Berkeley-based organization EcoEquity, as part of the Climate Equity Reference Project, has done the most detailed analyses of who owes what in the climate fight. He found that the United States would have to cut its emissions by 175 percent to make up for the damage it's already caused—a statistical impossibility. Therefore, the only way it can meet that burden is to help the rest of the world transition to clean energy and to help bear the costs that global warming has already produced. As Athanasiou put it, "The pressing work of decarbonization is only going to be embraced by the people of the Global South if it comes as part of a package that includes adaptation aid and disaster relief."

I said at the start that there is one sublime exception to the rule that we should be dousing fires, and that is the use of flame to control flame and to manage land—a skill developed over many millennia by the original inhabitants of much of the world. Of all the fires burning on Earth, none are more terrifying than the conflagrations that light the arid West, the Mediterranean, the eucalyptus forests of Australia, and the boreal woods of Siberia and the Canadian north. By last summer, blazes in Oregon and Washington and British Columbia were fouling the air across the continent in New York and New England. Smoke from fires in the Russian far north choked the sky above the North Pole.

For people in these regions, fire has become a scary psychological companion during the hot and dry months—and those months stretch out longer each year. The *San Francisco Chronicle* recently asked whether parts of California, once the nation's idyll, were now effectively uninhabitable. In Siberia, even last winter's icy cold was not enough to blot out the blazes; researchers reported "zombie fires" smoking and smoldering beneath feet of snow. There's no question

> "In Siberia, even last winter's icy cold was not enough to blot out the blazes; researchers reported 'zombie fires' smoldering beneath feet of snow."

that the climate crisis is driving these great blazes—and also being driven by them, since they put huge clouds of carbon into the air.

There's also little question, at least in the West, that the fires, though sparked by our new climate, feed on an accumulation of fuel left there by a century of a strict policy which treated any fire as a threat to be extinguished immediately. That policy ignored millennia of Indigenous experience using fire as a tool, an experience now suddenly in great demand. Indigenous people around the world have been at the forefront of the climate movement, and they have often been skilled early adopters of renewable energy. But they have also, in the past, been able to use fire to fight fire: to burn when the risk is low, in an effort to manage landscapes for safety and for productivity.

Frank Lake, a descendant of the Karuk tribe indigenous to what is now northern California, works as a research ecologist at the US Forest Service, and he is helping to recover this old and useful technology. He described a controlled burn in the autumn of 2015 near his house on the Klamath River. "I have legacy acorn trees on my property," he said—meaning the great oaks that provided food for tribal people in ages past—but those trees were hemmed in by fast-growing shrubs. "So we had twenty-something fire personnel there that day, and they had their equipment, and they laid hose. And I gave the operational briefing.

"I said, 'We're going to be burning today to reduce hazardous fuels. And also so we can gather acorns more easily, without the undergrowth, and the pests attacking the trees.' My wife was there and my five-year-old son and my three-year-old daughter. And I lit a branch from a lightning-struck sugar pine—it conveys its medicine from the lightning—and with that I lit everyone's drip torches, and then they went to work burning. My son got to walk hand-in-hand down the fire line with the burn boss."

Lake's work at the Forest Service involves helping tribes burn again. It's not always easy; some have been so decimated by the colonial experience that they've lost their traditions. "Maybe they have two or three generations that haven't been allowed to burn," he said. There are important pockets of residual knowledge, often among elders, but they can be reluctant to share that knowledge with others, Lake told me, "fearful that it will be co-opted and that they'll be kept out of the leadership and decision-making." But, for half a decade, the Indigenous Peoples Burning Network—organized by various tribes, the Nature Conservancy, and government agencies, including the Forest Service—has slowly been expanding across the country.

There are outposts in Oregon, Minnesota, New Mexico, and in other parts of the world. Lake has travelled to Australia to learn from aboriginal practitioners. "It's family-based burning. The kids get a Bic lighter and burn a little patch of eucalyptus. The teenagers a bigger area, adults much bigger swaths. I just saw it all unfold." As that knowledge and confidence is recovered, it's possible to imagine a world in which we've turned off most of the man-made fires, and Indigenous people teach the rest of us to use fire as the important force it was when we first discovered it.

Amy Cardinal Christianson, who works for the Canadian equivalent of the Forest Service, is a member of the Métis Nation. Her family kept trapping lines near Fort McMurray, in northern Alberta, but left them for the city because the development of the vast tar-sands complex overwhelmed the landscape. (That's the 173 billion barrels that Justin Trudeau says no country would leave in the ground—a pool of carbon so vast the climate scientist James Hansen said that pumping it from the ground would mean "game over for the climate.") The industrial fires it stoked have helped heat the Earth, and one result was a truly terrifying forest fire that overtook Fort McMurray in 2016, after a stretch of unseasonably high temperatures. The blaze forced the evacuation of eighty-eight thousand people and became the costliest disaster in Canadian history.

"What we're seeing now is bad fire," Christianson said. "When we talk about returning fire to the landscape, we're talking about good fire. I heard an elder describe it once as fire you could walk next to, fire of a low intensity." Fire that builds a mosaic of landscapes that, in turn, act as natural firebreaks against devastating blazes; fire that opens meadows where wildlife can flourish. "Fire is a kind of medicine for the land. And it lets you carry out your culture—like, why you are in the world, basically."

Chapter Two

The Realities of the Climate Imperative

BF Nagy

In his book *Wandering Home*, Bill McKibben describes a walk through the forests and fields of Vermont and New York's Adirondacks and writes: "What I've done, in my daily life and my political work and my writing, I've done because of these woods, these very woods we're walking through. I fell in love with these hemlocks, these steep slopes, these patches of rock, these streams lit by leaf-filtered sun." Obviously a talented writer, economist, a lover of the outdoors and community, as McKibben meandered through the hills and valleys, he also pondered the practical benefits of nature, its esthetic value, and our need to preserve its riches.

"Peering out through the branches, I watched dragonflies float above the lazy river, and listened to the rising tremolo pulse of insect song . . . the walls are filled with little streaks and swirls and flickers that please the eye like the dance of flames in a fireplace. Before long you're beginning to think in other ways that used to be heresy—like, why does my floor have to be all one type of wood, a big expanse of unbroken oak? Why can't it be like the forest that surrounds us, which is roughly equal parts birch and beech and maple?" His practical approach to day-to-day living in harmony with nature and his understanding of commonsense economics have made McKibben immensely popular.

His metaphorical delight for flame in the foregoing connects to the assertions in his article about the relationship we have with fire. Like stars and planets in the sky, the mystery and romance of flames and sparks from a late-night bonfire transform us back into enthralled children, who are drawn close, perhaps underestimating its perilous power.

Before learning to gain advantage with fire, ancient humans were likely afraid of it, after lightning-induced blazes killed or severely burned people and animals or destroyed areas in which we lived. As McKibben notes,

21

eventually we learned to control it for warmth, light in the evening, protection from predators, cooking food, and making tools. It was the gift that changed everything.

And we were undoubtedly also captured by its magnetism, beauty, and magic; families would soon gather around the fire, as we still do during campsite singalongs or in front of cozy hearths, staring for long periods into the flickering golden flames. It's irresistible. If we could stare at the sun without going blind, we would do so.

COMBUSTION

Combustion, too, the modern version of fire, has fascinated humans for centuries; but we are beginning to realize that devices and vehicles that burn oil, coal, and gas, nuclear power, nuclear bombs may be the equivalent of the mythological Icarus flying too close to the sun. Like McKibben, most of us view the climate crisis in the context of ordinary pragmatic realities: happy and healthy families, making a living, and securing our future. It is now clear to a vast majority of humans, more than 70 percent in most countries,[1] and up to 97 percent of scientists,[2] that combustion technologies have become obsolete. It's time to reduce risk, before high temperatures melt us completely and lead to our collective demise.

Much of McKibben's recent activism reflects his understanding that to make change quickly, we have to follow the money, appeal to the best instincts of powerful people. The economics of the energy transition are a key part of the current public discourse, and the economic benefits of clean energy and clean water are compelling.

MODERN BUSINESS SUCCESS

The most obvious current business opportunities include riding the massive wave of clean technology disruptions, introducing new products and brands, saving operational and supply chain costs on buildings, data centers, or large transport fleets; capitalizing on efficiency leaps (heat pumps, robotics, etc.); or attracting investment, based on emerging government intervention (e.g., Inflation Reduction Act).

Companies might also avoid uninsurable, stranded assets and margin erosion through diversification away from fossil fuels and combustion engines or gas boilers. You could work in new product categories (e.g., sharing economy, electric buses, renewables software, etc.); or improve underperforming technologies for new global needs (e.g., industrial electrification, desalination, or sustainable aviation).

For decades, Harvard Business School,[3] the *Economist*,[4] and the *Wall Street Journal*[5] have been covering a continuous stream of studies and business developments that verify that incorporating sustainability, employee health supports, and other forms of corporate responsibility achieves better performance in terms of earnings, revenue, outlook, growth, investment, share price, productivity, talent retention, innovation, and employee motivation. The most recent studies suggest that Millennials and Gen Z choose to work for and buy from ethical companies.[6]

> "North America seems to have suddenly realized that the billionaires of 2040 will all be in the climate solutions business. Every single one."

Avoiding support for climate solutions represents a business strategy that's so poor that anyone suggesting it should be terminated from their executive position. In some companies, there are still some who think going green means making a symbolic gesture toward sustainability and then issuing a press release. My advice: retire now while it's still your decision.

HOW COUNTRIES AND BUSINESSES CAN WIN

In addition to the industrial base that North America offshored to Asia, China, India, and Asia have bitten off a good chunk of the digital economy, and in the past decade, they have positioned themselves well for the shift to clean energy and water. Now in the 2020s, North America seems to have suddenly realized that the billionaires of 2040 will all be in the climate solutions business. Every single one. Clean energy activity is now at a fever pitch, and businesses and countries who wish to succeed had better get on board now.[7]

If employees steal equipment, damage property, or drive fleet vehicles recklessly, as competent business managers, we often terminate their employment because they have become threats to company assets. Yet we sometimes talk about sustainability as if it's for weed-smoking, green wackos. It's not. It's for analysts who can increase your plant efficiency by 20 percent and cut operating costs by 25 percent.[8] These people are not lightweights, and our home planet, our shared wealth, is not disposable.

The planet is the key asset on the balance sheet of Corporate Earth. We express shock if the government wishes to tax, fine, or sue us when we perpetrate damage to the asset. Let's stop being hypocrites. Let's protect our key asset at least as well as we protect a stapler that some misguided team member takes home from the office, or a pair of gloves from the plant.

Is the world really going green, or is this just wishful thinking by environmentalists? For many of us, our worldview is affected by information coming from the internet, television, and radio. And although news providers try to make their offerings seem unbiased, they're not. News is a product. Clicks or ratings create media revenue. News is managed as much by marketing managers as by editors. We might think everyone is reading or hearing the same news, but that's not happening either. Online marketers use algorithms to tell each of us what we want to hear. If we linger on something, it becomes part of our customer profile, and the algorithm keeps giving us more, continuously narrowing our individualized echo chamber.

> "News is managed as much by marketing managers as by editors. I've condensed thirty days of news from my social media feed into a few paragraphs. Some of it you may not have heard before."

As an environmentalist, I might receive very little of the same news that is received by people who have different interests. I've condensed thirty days of news from my social media feed into a few paragraphs. Some of it you may not have heard before:

Electric Vehicles—The International Energy Agency says 2023 electric vehicle (EV) sales will increase by 35 percent to fourteen million. Global EV sales were 1.26 million in June, up 38 percent compared to a year earlier. EV sales are up to about 26 percent of new car sales in California, the world's sixth-biggest economy. In California, Tesla now outsells Toyota. The first Cybertruck has come off the line at Tesla Giga Texas. Ford is tripling production capacity to satisfy soaring demand for electric F-150 Lightning pickup trucks. Seven big car companies have joined forces to add thirty thousand fast chargers across America, powered by renewables. Tesla is now opening 3.5 supercharger stations per day. Ford, General Motors (GM), and six other car companies have adopted Tesla NACS charger as the world standard. New York and seven other states will phase out emission cars by 2035. Canada, too. Avis and Uber are purchasing large Tesla fleets for the United States and Japan. BYD in China made one million electric cars in its first thirteen years, then three million in eighteen months, then five million in nine months. Tesla is in talks to enter the market in India, the world's most populous nation. Volkswagen increased EV sales 50 percent in the first half of the year, including Europe at 68 percent and the United States at 79 percent. A year after it opened, Tesla's Berlin plant will already double production. EV sales are up 100 percent in Australia, 65 percent in Ireland, and 60 percent in Sweden.

There are six hundred thousand EVs on the road in Norway, where about 90 percent of new car sales are electric.

Manufacturing—The Inflation Reduction Act helps land a one-billion-dollar solar factory in New Mexico. The Department of Energy funds a new battery factory in St. Louis. Other announcements: a new solar factory in Minnesota, new electric bus plant in Illinois, new Colorado solar cell factory with two-gigawatt capacity. A new lithium refinery in the United Kingdom will power about one million electric vehicles. A UK gas boiler maker switches to heat pumps.

Solar and Renewable Power—A report shows that the United States has 16.4 billion square feet of space on warehouse rooftops that will work for solar. A warehouse solar parking lot in Austin supplies 100 percent of the electricity needed for the building. Twenty-five states do not allow NIMBYs in homeowner associations to disallow rooftop solar. At twenty-two megawatts, the world's biggest rooftop solar install is complete in Thailand. Disney boosts revenue by installing sixty-seven thousand solar carports at its Paris theme park. A Mexico City food market has thirty-five thousand solar panels on its roof. In Europe, there are now about ten million residential rooftop solar arrays and 1.1 million residential home batteries.

The world will add 440 gigawatts of renewables this year, about 30 percent more than last year. In a single year, Denmark makes fourteen billion in profits from wind power. A Bloomberg report says nuclear and hydrogen cost $200+ per megawatt/hour, gas and coal cost around $125, utility solar under $50. Renewables will become the largest electricity source next year, exceeding contributions from coal, gas, nuclear, and oil. Walmart, Meta, GM, and other big companies are building solar, wind power, and now smart transmission infrastructure. America's biggest utility drops membership in the American Gas Association lobby group. Ad and PR agencies pledge to stop helping fossil fuel companies spread disinformation. Iowa wind power record—64 percent of electricity production. Study: solar plus agriculture increases farmland value by 30 percent. Tesla will install $413 million in battery megapacks in Massachusetts. To keep a crumbling, old California nuclear plant running could cost ratepayers forty-five billion dollars, which could buy twenty times as much renewables instead. The World Economic Forum says following the Ukraine invasion Europe replaced gas plants with solar, saving twenty-nine billion dollars, but France is wasting twenty-eight billion dollars to keep overpriced, dysfunctional nuclear plants operating. In the first six months of 2023, Germany installed solar farms equivalent to seven nuclear plants. There was one death and more than one hundred hospitalized when a Russian nuclear plant exploded. An Australian Minister says 82 percent renewables can be met. Mexico builds a 18,990-panel solar farm that will power 1.3 million homes. Spain is at 42 percent renewables and is targeting 81 percent. A new Scottish wind farm will power five thousand homes and employ six thousand people. A Hungarian nuclear plant cuts power because of warm Danube River waters. For the second consecutive summer, many of France's nuclear plants shut down or limit output during hot weather. All trains in the Netherlands run on renewables.

Policy—Boston has banned fossil fuels in municipal buildings. California is adding massive grid backup, requires all new buildings to include solar panels, all public buses to be zero-emissions. Maine approves offshore wind power for 50 percent of its electricity. Massachusetts blows past the one hundred thousand heat pump target, increases the goal to 175,000. Canada announces phaseouts of fossil fuel subsidies. Brazil's returning President Lula cut Amazon rainforest destruction by 60 percent, introduced 100+ other climate measures, attracting global investors. Lower Saxony government in Germany tested fourteen hydrogen trains but is switching to electric because of high cost.

Extreme Weather—Swiss Re, one of the globe's biggest reinsurers, warns extreme weather event costs to increase by billions. Last year, there were eighteen one-billion-dollar-plus weather events, including Hurricane Ian at $114 billion. US heat records: more than five thousand broken in past thirty days. Seven hottest days in one hundred thousand years all happened in one week in July. Ninety million Americans under dangerous heat alerts. Antarctica: 2.6 million square kilometers of sea ice melted. Ocean heating temperatures have increased far faster than expected. In Arizona, people are falling down on 180°F pavement and ending up needing surgery for burns. Hawaii: eighty people dead from a firestorm that leveled a town in Maui. Most of Florida's tourist-attracting coral reefs are bleaching and dying. Heat wave in the Southwest: Death Valley hits 125°F. Phoenix reaches eighteen straight days above 110°F. Water levels in the Mississippi and Ohio Rivers fall for second straight year. Canada: thirty-two million acres of forest burned so far this year. A summer's worth of rain fell in twenty-four hours in Nova Scotia.

The temperature in China hit 126°F on July 16. Floods in Beijing after twenty inches of rain in twenty-four hours (seven inches the next day), eight hundred thousand people displaced. Floods in China's main grain-producing region kill fourteen, increase food prices. India banned grain exports due to crop damage. Food exporters Italy and Ukraine are struggling, too. Wildfires caused by climate change detonate ammo at Greek NATO airbase, resulting in evacuation of nearby towns. Thousands flee wildfires on Greek island of Rhodes. Heat wave red alerts issued for sixteen cities in Italy. Mediterranean wildfires spread to Portugal and Croatia "like a blowtorch." Slovenia floods kill three, cause millions in damages. Homes flow downriver in Norway flood. Extreme floods and landslides kill more than one hundred people in Pradesh, India. Iran shuts down its economy for two days due to "unprecedented heat." Floods in Morocco. Floods in Pakistan. Drought leaves millions in Uruguay without drinking water. The fossil fuel industry gets eleven million dollars every single minute in subsidies. Vandana Shiva: "It's not an investment if it's destroying the planet." Meehan Crist: "The economy is a wholly owned subsidiary of nature."

THIRD ACT

In August of 1969, during "the Summer of Love," half a million young people in their friendship bracelets and tie-dye clothing painted peace signs and monarch butterflies on their skin. They packed their flutes, flowers, and ukuleles and headed for Max Yasgur's dairy farm near Bethel, New York, for Woodstock, an "Aquarian exposition."

Peace and brotherly love were more than a meme or a movie concept in those days. It was a statement of defiance, chanted by a generation at the epicenter of a decade of anti–Vietnam War marches and race riots in America. President Kennedy and Martin Luther King Jr. were shot dead, and four students at Kent State soon would be. It was widespread pressure from ordinary citizens against a powerful business and government elite that resulted in President Lyndon B. Johnson and Prime Minister Pierre Trudeau passing legislation that sought to end racism, sexism, and agcism that had pervaded post-war North America. Many at Woodstock had hoped that a new era was dawning and that we would begin healing.

Hundreds of thousands of music lovers left their anger in the cities and reveled in the lush bluegrass, milkweeds, and oxeye daisies, shrugged off the shortage of food and sanitation and went skinny-dipping in the Filippini Pond behind the Woodstock stage. Even in the rain and mud, they sang together, and on the third day, their host, Max Yasgur, said, "You've proven to the world that . . . a half million young people can get together and have three days of fun and music and have nothing but fun and music."

Baby boomers left behind the sweet scent of hyacinth, reeds, and ferns, the birdsong from blue jays and yellow warblers (and Janis Joplin, Neil Young, Joan Baez) and moved through the workaday times in their lives, having families, creating communities, and building careers. Did we remain true to the peace, love, and understanding of the Woodstock garden? The pressures of our mortgages, fears, and ambitions changed us, and we've committed the sins we most loudly railed against: materialism and apathy. After improving

> "After improving for a while, the world has begun, on our watch, to regress, not to advance. We've handed it over to the worst among us, the profit-takers."

for a time, the world has begun, on our watch, to regress, not to advance. We've handed it over to the worst among us, the profit-takers, who are leading us to ruin, and we have no one to blame but ourselves. The story of climate change is the story of the social justice generation losing its way.

Today, through his non-profit Third Act, Bill McKibben is offering us one last chance to leave the world better than we found it. It's time to send a new message of defiance to those in charge, rally courageously once again, take back control, and usher in the dawning of the Age of Aquarius.

ECONOMICS AND FATALITIES

Professor Mark Jacobson explains global clean energy from economic and scientific perspectives. Key findings of his Stanford team are that the clean energy transition is accelerating and will pay for itself quickly, in just six years.[9]

This short payback will be achieved because of dramatic system efficiency improvements that come with clean renewable energy, batteries, and intelligent systems. Our investment will be recovered quickly, and the new system will keep saving us more after we hit breakeven, at a rate of 100 percent in six years, or about 16 percent annually. In addition, we will save billions in property damage and uninsurable asset losses, and as noted by McKibben, we will also save millions of lives.

> "Jacobson's rallying cry 'Deploy, deploy, deploy' is the tagline that we should be shouting from the rooftops."

The shift to clean, renewable energy offers numerous economic and non-economic benefits to humanity and is now unstoppable. However, despite the fact that we are heading in the right directions and ramping up good solutions, we have fallen behind in the race to mitigate some of the worst effects of climate change. As mentioned, millions are now dying from air pollution each year and millions from related crises like intensified pandemics. Thousands are dying from extreme weather disasters, and scientists say these will soon be killing millions. Jacobson's rallying cry "Deploy, deploy, deploy" is the tagline that we should be shouting from the rooftops.

Chapter Three

100% Renewables and the Six-Year Payback

Professor Mark Z. Jacobson

Professor **Mark Z. Jacobson** and his Stanford research group are well known for advising the Bush, Obama, and Biden administrations in the White House, the US Congress, several state governments, and beyond. They have provided renewables roadmaps for fifty-three towns and cities,[1] seventy-four metropolitan areas,[2] all fifty US states, and 139 countries.[3] Professor Jacobson has addressed the top international climate conferences, founded important educational groups, and created high-impact campaigns with celebrities like Leonardo DiCaprio, Mark Ruffalo, and Leilani Münter. He even appeared on the *Late Show with David Letterman*. More biographical information can be found on Professor Jacobson near the end of this book.

Do we need technologies such as carbon capture, direct air capture, bioenergy, and nuclear power to reduce climate-damaging emissions? Two new studies add to an already-abundant body of similar research that confirms that we can create a stable power grid in the United States and 144 other countries using clean, renewable electricity and heat; batteries; and other available non-emitting energy technologies. The studies demonstrate that up-front investment may be recovered in just six years because the cost and energy required to operate the new system is considerably reduced relative to a business-as-usual system. We can create more jobs and reduce climate-damaging emissions, including carbon and methane, without blackouts, at low cost, and eliminate over fifty-three thousand US air-pollution-related deaths and millions more illnesses each year. Furthermore, carbon capture, direct air capture, bioenergy, and nuclear power are unhelpful distractions that have already resulted in significant unnecessary opportunity costs.[4]

The key energy-use sectors are electricity, buildings, transport, and industry. For electricity, the main clean, renewable, electricity-generating technologies are wind, solar and concentrated solar, geothermal electricity, hydroelectricity, and tidal and wave electricity. Collectively, these are Wind-Water-Solar (WWS) technologies. All but tidal and wave are fully commercialized. In fact, wind and solar are currently the lowest-cost electricity-generating technologies, according to Lazard, a financial services firm which has become the established expert source for levelized energy cost analysis.[5]

> "Due to the approximately $11 trillion in annual cost savings, the payback time for the new energy system is less than six years."

Two new studies from my research group at Stanford University conclude that these problems can be solved in each of the fifty United States and in 145 countries we examined, respectively, without blackouts and at low cost, using almost all existing technologies.[6] We and over thirty other research groups have shown that, with clean renewables alone, this transition reduces energy costs and land requirements while creating jobs.[7]

Worldwide, the energy that people actually use will decrease by more than 56 percent with an all-electric system. The up-front cost to replace all energy in the 145 countries, which emit 99.7 percent of the world's human-produced carbon dioxide, is about $62 trillion. Due to the approximately $11 trillion in annual energy cost savings, the payback time for the new system is less than six years.[8]

EFFICIENCY AND JOB CREATION

There are five reasons for the worldwide energy-use reduction of about 56 percent using clean, renewable sources: the efficiency of electric vehicles (EVs) compared with combustion vehicles, the efficiency of electric heat pumps for air and water heating compared with combustion heaters, the efficiency of electrified industry, the elimination of energy needed to obtain fossil fuels, and miscellaneous additional efficiency improvements.

The new system will also create as many as twenty-eight million more long-term, full-time jobs than will be lost worldwide and require only about 0.53 percent of the world's land for new energy, with most of this area being empty space between wind turbines on land. Such space can be used for multiple purposes. Thus, we found that the new system requires less energy, costs less, and creates more jobs than the current system.[9]

RELIABLE POWER GRID

Our research also shows that we have abundant wind and solar potential, and wind and solar can each power the world's all-purpose energy many times over. We also found that no batteries with more than four hours of storage are needed. Instead, long-duration storage can be obtained by concatenating batteries with four-hour storage together. We also found that even if battery prices are 50 percent higher than projected in 2035, overall energy costs will be only 3.2 percent higher than our base estimate. However, as of 2023, battery costs are still on a decreasing trend, and performance from the newest batteries is increasing rapidly.[10]

Clean, renewable energy is naturally replenishable, results in no emissions of health- or climate-affecting air pollutants, and does not pose other major environmental threats. Risk analysts in government, business, and financial organizations have been interested in our estimates showing that, in the United

> "The more that storage costs decline, the more storage will be coupled with wind and solar to keep the grid stable."

States, a full transition to WWS in 2050 may reduce energy needs by more than 56 percent, energy costs by approximately 63 percent, and energy-plus-health-plus-climate costs by about 86 percent—compared with a business-as-usual scenario.

Wind and solar energies are complementary insofar as the sun often shines when the wind isn't blowing, and vice versa. Combining wind and solar smoothens the power supply.

Similarly, combining wind or solar energy among distant facilities can also help reduce variability in either, compared with at one location. Locating wind turbines in coastal waters helps meet spikes in electricity demand because offshore wind is usually less variable than is onshore wind and often peaks when electricity demand peaks.[11]

Building more wind turbines in cold climates increases reliability because, on average, when temperatures drop and heating demand goes up, winds become stronger.

Gaps in wind and storage supply can also be mitigated through the use of batteries and other kinds of electricity storage, such as conventional hydropower, pumped hydropower, flywheels, compressed air storage, and hydrogen fuel cells. In many places, solar plus batteries was already less expensive in 2022 than coal or nuclear and is replacing both.[12] Battery costs have declined 97 percent since 1991. The more that storage costs decline, the more storage will be coupled with wind and solar to keep the grid stable.

A broad spectrum of energy conservation measures can significantly reduce energy use and contribute to increased grid stability.

In 2009, Dr. Mark A. Delucchi, a research scientist at the Institute of Transportation Studies at the University of California, Davis (and later at Berkeley), and I prepared an article for *Scientific American* based on research conducted by our teams.[13] We noted that according to the US Energy Information Administration, the annual-average power consumed worldwide was about 12.5 terawatts. The agency projected that, in 2030, the world might require 16.9 terawatts of average power as global population and living standards rise, with about 2.8 terawatts in the United States.

The sources of that energy depended heavily on fossil fuels, but if the planet were powered entirely by WWS, with no fossil fuel or biomass combustion, an intriguing level of savings might occur. Global power demand might decrease to only 11.5 terawatts, and US demand would decrease to 1.8 terawatts. That decline occurs because, as mentioned, electrification is a more efficient way to use energy than is combustion. For example, only 17 to 20 percent of the energy in gasoline is used to move a vehicle (the rest is wasted as heat), whereas 75 to 86 percent of the electricity delivered to an electric vehicle goes into motion.

> "Only 17 to 20 percent of the energy in gasoline is used to move a vehicle whereas 75 to 86 percent of the electricity delivered to an electric vehicle goes into motion."

Even if demand after electrification did rise to 16.9 terawatts, WWS sources could provide that additional power and more. Detailed studies by us and others indicated that the maximum possible energy extracted from the wind, worldwide, is about 250 terawatts. Solar, alone, is about 19,400 terawatts. Wind and sun out in the open seas, over high mountains, and across protected regions would be less available. If we subtract these and low-wind areas that are not likely to be developed, we are still left with forty to eighty-five terawatts for wind and 1,350 terawatts for solar, each far beyond future human demand.

The other WWS technologies will help create a flexible range of options. Although all the sources can expand, for practical reasons, wave power can be extracted only near coastal areas. Many geothermal sources are too deep to be tapped economically. And even though hydroelectric power now exceeds all other WWS sources, most of the suitable large reservoirs are already in use.

MOMENTUM

In nine countries, 98.5 to 100 percent WWS generation was already a reality by 2021. The states of South Dakota, Vermont, Washington, Idaho, Iowa, California, and Texas, along with the countries of Scotland and Germany, will soon reach this milestone. It will require a great deal more electricity generation from wind and solar.

Do we need direct air capture to reduce CO_2 already in the atmosphere? No. We can limit CO_2 in the air to 350 parts per million (ppm) by stopping 80 percent of emissions by 2030 and 100 percent by no later than 2050.[14] Removing carbon dioxide from the air with equipment (direct air capture), as opposed to with trees, is an opportunity cost, so it is always better to invest in a different mitigation method.

Similarly, removing carbon dioxide from the exhaust stacks of fossil power plants and storing the CO_2 underground (carbon capture and sequestration) is an opportunity cost. Even if it is powered by renewable electricity, it requires so much electricity that the same electricity can be used more efficiently simply to replace the fossil fuel power plant, thus reducing carbon dioxide more. In addition, by replacing the fossil plant, renewable electricity eliminates the air pollution, mining, and infrastructure associated with the fossil plant. Carbon capture does none of that. As a result, it is always far better to use renewable electricity to replace a fossil plant than to run carbon capture for any purpose, including for direct air capture.

Nuclear electricity results in nine to thirty-seven times the carbon dioxide–equivalent emissions per unit energy than does wind electricity, when reactor construction and uranium refining and transport are considered. At present, nuclear electricity generation costs five to ten times per unit energy that of new wind and solar, and the planning, approvals,

> "Very little new technology is needed, particularly not carbon capture, direct air capture, modern bioenergy, or modern nuclear power."

and construction of new nuclear plants require ten to twenty-two years to complete (versus one to three years for new wind or solar), frequently exceeding original budgets by billions of dollars. In addition, nuclear power generation comes with significant waste disposal and terrorism risks.[15]

We already have 95 percent of the technologies we need today to address both energy and non-energy emissions. Very little new technology is needed, particularly not carbon capture, direct air capture, modern bioenergy, or modern nuclear power.

Buildings—Do we need new technologies to clean up energy in buildings? Emissions from buildings arise from natural gas use for space heating, water heating, cooking, and clothes drying. Electric replacements for these are commercially available with dropping costs. They include electric heat pump air heaters and air conditioners, electric heat pump water heaters and clothes dryers, and electric induction cooktop stoves. Heat pumps use about one-quarter the energy as gas heaters, thereby costing less than gas heaters over their useful lives.

Transportation—Electric vehicles are now commercialized, and their global adoption is increasing quickly. They are replacing fossil-fuel vehicles of all types and weights, apart from long-distance aircraft and ships, the longest-distance trucks and trains, and heavy military vehicles. Long-distance, heavy vehicles are part of the last 5 percent of energy technologies that may take until 2025 or 2035 to commercialize. These vehicles may ultimately operate using fuel cells powered by green hydrogen from renewable electricity. No biofuels, such as ethanol, biodiesel, or bio-jet fuel, are needed.

Cement, Steel, and Industry Processes—Heat for high-temperature processes will come from existing electric technologies: electric arc furnaces, induction furnaces, resistance furnaces, dielectric heaters, electron beam heaters, heat pumps, and CSP steam.[16]

Steel and cement manufacturing emit CO_2 from chemical reaction in addition to from energy. Biomass burning emits CO_2, black and brown carbon (second-leading cause of global warming), and more. Rice paddies, landfills, and manure emit methane. Halogens, used as refrigerants and solvents, leak. Fertilizers emit nitrous oxide. These all have current solutions. For steel, the hydrogen direct reduction process combined with heat powered by WWS reduces CO_2 during steel production by 97.2 percent.[17]

Replacing concrete with geopolymer concrete or Ferrock eliminates chemically produced CO_2 from concrete production.[18] Recycling concrete helps further. Strong policies are needed to phase out biomass burning. Collecting methane from rice paddies and landfills with pipes is an existing technology, as is a digester for trapping methane from manure. Using less nitrogen-based fertilizer and cultivating leguminous crops that don't require fertilizers help reduce nitrous oxide emissions.

Some argue that we need to capture the chemically produced CO_2 from concrete. Instead, it can be reduced using geopolymer concrete or Ferrock and recycling concrete. Adding carbon capture to existing concrete production facilities is not a good option. It is an opportunity cost. If the same investment for the capture equipment and renewable electricity used to power the capture equipment is used instead to replace a fossil fuel electricity or heat source, it reduces more CO_2 and simultaneously eliminates air pollution and mining.[19]

Further, it is more efficient to invest in reducing biomass burning, eliminating both CO_2 and air pollution, or to increase reforestation, than to invest in carbon capture or direct air capture. As of 2023, captured carbon is still used mostly to enhance oil extraction and create polluting synthetic fuels for burning.

BIOMASS AND ETHANOL

In addition, biomass used for heat and ethanol used for transportation are currently ineffective as climate solutions. Blue hydrogen (hydrogen produced from natural gas with carbon capture equipment added) is not a climate solution.[20]

One study calculated that biomass burning causes a twenty-year-average global warming of about 0.4 Kelvin when accounting for heat and moisture fluxes, all major gas and aerosol constituents (including black and brown carbon and tar balls, among others), cloud absorption effects, semidirect effects, and indirect effects. Biomass burning is estimated to cause approximately 250,000 premature deaths each year, with more than 90 percent due to particulate matter, and the remainder due to ozone.[21]

"Biomass used for heat and ethanol used for transportation are currently ineffective as climate solutions. Blue hydrogen is not a climate solution."

When burned in vehicles, even the most ecologically acceptable sources of ethanol create air pollution that will cause the same mortality level as when gasoline is burned. Due to the energy intensity of and land required for producing ethanol, it results in similar CO_2-equivalent emissions as gasoline, plus it creates negative environmental impacts on water and air.

CHALLENGES WITH ALL-OF-THE-ABOVE ENERGY STRATEGIES

Sometimes high-profile people who want to maximize positive reviews for their public statements will say they support an all-of-the-above energy strategy, which generally means Wind-Water-Solar (WWS) technologies, plus nuclear, fossil fuels with carbon capture, direct air capture, bioenergy, and electrofuels. However, as explained in this chapter, some of these technologies constitute unacceptable opportunity costs, attracting significant investment while resulting in either more air pollution, more greenhouse gas emissions, more energy-security risk, more land-use risk, higher costs, longer planning-to-operation times, or all of these, relative to WWS.

An all-of-the-above strategy is wasteful, costly, and slow. It provides reduced benefit and a longer delay before we eliminate air pollution, global warming, and energy insecurity. Given the increase in ocean temperatures, rapid glacier melt, and numerous other indicators, we no longer have time for misplaced priorities and experimentation. It is better to focus on solutions we know work that can be implemented on a large scale quickly and at low cost.

By using only clean, renewable wind, water, and solar energy generation and storage, we can address climate, energy insecurity, and seven million annual air pollution deaths worldwide together. The key is to deploy, deploy, deploy existing clean, renewable, safe energy as fast as possible.

Chapter Four

Selling Our Home to the Highest Bidder

BF Nagy

When it comes to climate, some businesspeople or politicians need greater clarity and commitment. Others deliberately create or amplify disinformation due to self-interest such as next quarter's earnings or next year's election. The latter are usually fossil fuel companies, or are bankrolled by them, and their massive disinformation campaigns do an incalculable disservice to humanity. People will suffer and die due to their misleading messaging.

DISINFORMATION AND DARK POLITICS

Details on the misdeeds of fossil fuel and nuclear companies can be found in many places, for example in *The New Climate War*, a book by Dr. Michael E. Mann, distinguished Nobel prize–winning professor at the University of Pennsylvania.[1] Environmentalists have tried to help these energy players and guide them to clean, profitable solutions. But rather than protect their children, grandchildren, and shareholders by shifting to modern ways, they have chosen to confuse everyone.

It is incomprehensible that these firms, managed by human beings, would choose this course of action. In the early 2020s, they were at least greenwashing, which seemed like an early step toward actually changing. Then they reversed and seem to have chosen to drive their firms as is until they die, like an old car that's not worth fixing. This is strategically devoid of inspiration and morally subterranean.

FALSE SOLUTIONS

No matter how credible an argument sounds, if it suggests we should keep using fossil fuels, that should be a red flag. If it says it will cost us more to switch to clean renewables, or they don't work, it is false. The idea that we should delay clean energy is dangerous. Requests for more research money for the fossil industry have lost all credibility. If it criticizes a peer-reviewed scientific study published in a credible journal and its criticism does not appear in the same journal, the writer generally has been paid by fossil fuel firms to create bogus arguments.

After fossil fuel companies loudly declared for years that carbon capture could solve all our problems, billions of dollars were allocated under the Inflation Reduction Act and the Environmental Protection Agency (EPA) provided funds with emissions-tracking guidelines. Once their bluff had been called and the funding was conditional on measured results, fossil fuel companies declared the program was badly designed. Trapped by their own dishonesty, they backtracked, admitting the technology couldn't achieve their prior claims.[2] More than forty projects in the world have been cancelled, according to the Massachusetts Institute of Technology (MIT), each costing billions in public money.[3]

> "If it says it will cost us more to switch to clean renewables, or they don't work, it is false."

In 2023, a handful of youths in Montana, as in numerous places around the world, sued the government for allowing pollution and climate inaction. They won. The judge agreed that the state had violated youths' constitutional right to a clean and healthy environment. Stanford scientist Professor Mark Jacobson testified without pay in favor of the young people. The state paid $95,000 in tax dollars for opposing witnesses who said pollution and climate inaction are okay. This perfectly encapsulates the insanity of the current climate disinformation war worldwide.

Back in 2011, Professors Robert Howarth and Anthony Ingraffea showed that shale gas was a greater source of CO_2 than coal. Howarth also made news with Professor Jacobson of Stanford in 2022, when they established that blue hydrogen is not a climate solution and that it actually adds to emissions.

In this book and in their many other pursuits, Howarth and Jacobson provide the science behind the consensus that carbon capture, hydrogen, and other technologies promoted by the fossil fuel industry as climate solutions are failed, disingenuous, theoretical, or unproven distractions. Even if they ever move beyond greenwashing to commercialization, they are unlikely to play a significant role in halting climate change. The unfortunate aspect of being brilliant, well-known scientists like Howarth, Jacobson, or Michael

Mann at University of Pennsylvania is that they spend an unfair proportion of their time not doing science and, instead, patiently clearing up disinformation about climate solutions that has been invented and obscenely amplified by profit-takers and careless media people.

In the 1980s, Howarth and fellow biogeochemist Roxanne Marino created a biogeochemistry seminar series at Cornell, which has been running continuously ever since. The wife-and-husband team also established the Howarth-Marino lab at Cornell, which has studied methane emissions from oil and gas, the impact of nitrogen pollution on aquatic ecosystems, harmful algal blooms, how climate, agriculture, and urbanism affect nutrient fluxes in major rivers around the world, and much more.

They have lived happily in Cornell's idyllic Ithaca area, enjoying some of the most famous examples of mother nature's artistry found anywhere, including at least fifteen babbling brooks with waterfalls such as Taughannock and Buttermilk, the Cornell Botanic Gardens, the Mulholland Wildflower Preserve, Watkins Glen State Park, the Gorge Trail, something like fifty vineyards, and the 750-acre Taughannock State Park. It offers hiking and nature trails, camping, swimming, fishing, and a marina on legendary Cayuga Lake.

The Cornell Botanic Gardens were established in 1875, ten years after the university was founded. The management team overseas fifteen waterfalls, an arboretum, and thirty-two miles of trails and bike paths, which attract about seventy thousand annual visitors. The arboretum is famous worldwide for being a home to beeches, hemlocks, sugar maples, pecan, hickory, and walnut and more than one thousand other species of trees and shrubs. Fall Creek, a defining feature of the Cornell campus, meanders through a stand of trees that have been here for more than three hundred years.

Says the Cornell Botanic Gardens web site: "By expanding access to the natural world, we can elevate our understanding of the interconnectedness of people and nature, and ultimately, help conserve our shared natural heritage." Not much of this has happened by accident, with the university and local governments continuously investing in and working at preserving and creating these national treasures. They have also built one of the world's top learning and research meccas, with Howarth and Marino included in their key assets. They, too, are national treasures. By contrast, fossil fuel companies continue to destroy the Earth in order to maximize profits.

SELLING TO THE HIGHEST BIDDER

First, the profit-takers in the oil and gas business perpetuate destruction of the environment, then they buy the influence of weak politicians who protect their profits and shell out obscene operating subsidies, utility price increases to finance expansion, and pointless research delays. This forces children into courtrooms to fight for the right to breathe and drink clean water because they're often arrested if they undertake peaceful free speech protests in the streets. Next, fossil companies use more public money to defend themselves in expensive, drawn-out lawsuits designed to delay change and systemize the slush flow. According to Reuters, big oil doubled its profits in 2022 to more than $219 billion.[4]

It's time to stop listening to fossil fuel and nuclear people and compromised politicians. Not only are they spending millions to spread disinformation,[5] they have inserted supporters seemingly on every energy board, government energy department, community group, task force, and standards committee. We think they are sincere neighbors. But apparently, they don't care if the Earth implodes, or your family is swept away in a firestorm. The red flags are in the fourth paragraph of this section. Please listen for the clues, and kick these people out of government and other groups.

The oil and gas business is a money-losing operation based on obsolete technology. Its profits are less than the public funds it receives. According to the IMF, government subsidies rose from more than $4 trillion in 2017 to about $7 trillion in 2022.[6] In many cases, fossil fuel company assets were also financed by the public. Weak politicians, useful media idiots, and online trolls publicly reveal their tenuous grasp of reality by describing oil and nuclear as free enterprise. More accurately, it's state-supported corporate welfare that extracts hard-earned money from ratepayers and displaces affordable, clean, healthy energy while the planet burns. It's time we decisively regulated or nationalized these companies and immediately phased out fossil fuels.

Chapter Five

False Solutions
from the Gas Industry

Professor Robert Howarth

Cornell's Professor **Robert Howarth** is an internationally renowned research scientist, professor of ecology and environmental biology, journal editor, and advisor to governments. *Time* named him in its "50 People Who Matter" 2011 Person of the Year issue. He was a participant at COP21, COP24, and COP26, and in the early 2020s, he worked on New York State's Climate Action Council. He has published over 220 scientific papers, reports, and book chapters, and his work has been cited more than eighty-two thousand times in the peer-reviewed literature, making him one of the most cited environmental scientists in the world. More information on Professor Howarth and his wife, biogeochemist research partner at Cornell Roxanne Marino, can be found near the end of this book.

Delay. Distract. For over fifty years, the oil and gas industry has followed this approach to marketing their products.[1] They have known full well that continued use of fossil fuels would have a devastating effect on our global climate system. And they have worked hard to discredit those who challenge their narratives and bring objective information to the public and decision-makers. The policy to delay and distract has worked for the gas industry: global consumption of gas has risen four-fold over the past fifty years.[2] In the United States, half of the gas ever burned has been produced in the past thirty-three years.[3] And 2022 saw record-high profits for the oil and gas industry.[4]

For the first decade or two of this twenty-first century, the strategy of the gas industry to delay and distract focused on gas as a "bridge fuel." The industry argued that the world was not yet ready to be powered just by renewable energy and that gas could be used to replace coal plants as a way to reduce greenhouse gas emissions. It worked in secret to gain allies from environmental groups, for instance, funding the Sierra Club campaign against coal, which turned into a pro-gas campaign.[5] Even President Barack Obama endorsed and promoted the bridge fuel concept.[6]

There was an element of truth behind this gas-as-bridge-fuel marketing: For the same amount of energy, the emissions of carbon dioxide from burning coal are substantially greater than from burning gas. Missed in this argument is the simple fact that the carbon dioxide emissions from burning gas are still substantial, and the use of any fossil fuel is incompatible with the goal of limiting global warming to 1.5°C from the pre-industrial baseline. Further, the bridge-fuel marketing completely ignored emissions of methane, an issue that I and co-authors Tony Ingraffea and Rene Santor raised in a 2011 peer-reviewed analysis of the greenhouse gas footprint of gas.[7]

Methane is an incredibly powerful greenhouse gas, responsible for more than eighty times as much warming as carbon dioxide per mass of gas emitted to the atmosphere for a twenty-year time period following emission.[8] According to the latest synthesis report from the Intergovernmental Panel on Climate Change, methane emissions have been responsible for 0.5°C of all global warming since the 1800s, compared to 0.75°C for carbon dioxide.[9]

> **"Methane is an incredibly powerful greenhouse gas."**

Methane cannot be ignored by anyone who cares about the climate. The combined warming consequences of these two gases exceeds the actual total warming to date since emissions of some other substances (principally sulfur dioxide) contribute to atmospheric cooling. No other greenhouse gases come close to carbon dioxide and methane when it comes to global warming.

Our 2011 paper was the first effort to examine how methane emissions contribute to the greenhouse gas consequences of shale gas, a type of gas that only became commercially possible in the twenty-first century due to the combined use of two new technologies, high-precision directional drilling and high-volume hydraulic fracturing ("fracking"). We tentatively concluded that because of methane emissions, the total greenhouse gas emissions from shale gas (and perhaps even conventional gas) might exceed those from coal, which directly challenged the bridge-fuel marketing of the gas industry.[10]

As an article in the *New York Times* stated on the day our paper was published in April 2011, "the findings are certain to stir debate. For much of the last decade, the gas industry has carefully cultivated a green reputation, often with the help of environmental groups that embrace the resource as a clean-burning 'bridge fuel' to a renewable energy future." The article went on to quote me: "I don't think this is the end of the story, . . . I think this is just the beginning of the story, and before governments and the industry push ahead on gas development, at the very least we ought to do a better job of making measurements."[11]

The pushback by the gas industry on our paper was immediate and intense, as described in a 2022 documentary from the BBC. For example, Melanie Kenderdine, who built her career in the gas industry before becoming Executive Director of the MIT Energy Initiative, was on CNBC TV the next morning, with very critical comments on our research. Her colleague at MIT, Ernie Moniz, soon became the Secretary of the US Department of Energy under President Barack Obama; throughout his time in government, Moniz promoted gas as a bridge fuel and regularly criticized me and my colleagues for our work on methane emissions. For two years following the publication of our paper, the American Gas Alliance paid for an advertisement on Google that was quite critical of me and my research, and that made many false allegations. This advertisement was the first hit anyone would find while searching for me on Google.

Science has vindicated our 2011 paper, which has now been cited 1,855 times in other peer-reviewed papers.[12] Literally hundreds of scientists have now measured and analyzed methane emissions from shale gas and conventional gas development and use, particularly in the United States. The preponderance of the data from these studies show methane emissions to be very similar to those we estimated in 2011, based on the far more limited set of data available then.[13] The most recent models show that even at very modest levels of methane emissions, the greenhouse gas consequences of using gas can be as bad or worse than burning coal.[14]

> "Science has vindicated our 2011 paper, which has now been cited 1,855 times in other peer-reviewed papers . . . hundreds of scientists have now measured and analyzed methane emissions from shale gas."

The bridge-fuel concept has lost its power as a marketing tool, and gas is now widely viewed as a false solution to the climate crisis. By the spring of 2021, even the relatively conservative International Energy Agency was signaling that we need to mandate immediate and sharp declines in the consumption of gas globally.[15] In the summer of 2021, the former head of the UN Framework Convention on Climate Change, Christiana Figueres, penned an editorial under the heading "Gas, like coal, has no future as the world wakes up to climate emergency."[16]

The gas industry has adapted by shifting its marketing campaigns away from gas as a bridge fuel and toward the promotion of hydrogen. Two major players in these efforts are FTI Consulting, a lobbying firm closely tied to the gas industry, and a group FTI helped establish in 2017 called the Hydrogen Council. Founding members of the Hydrogen Council include Aramco, Shell, BP, and TotalEnergies. Of interest, the contact address for the Hydrogen Council is the same as that for the European headquarters of FTI.[17] From its start, the Hydrogen Council has promoted using the existing gas pipeline delivery system for hydrogen to homes and commercial buildings, providing a reason to keep these highly profitable pipelines in operation even as consumption of gas falls.

"Blue hydrogen" has played a prominent role in the marketing by the Hydrogen Council and others. Currently, over 95 percent of the hydrogen produced in Europe and North America is made from gas. Gas is composed mostly of methane, and this methane serves as the feed product for a process invented well over one hundred years ago that converts the methane into hydrogen and carbon dioxide. The process uses high pressure and heat, and this heat energy comes from burning more gas. Greenhouse gas emissions from this hydrogen, called "gray" hydrogen, are huge.[18] With the invention of blue hydrogen, the gas industry claimed to be able to make hydrogen with near-zero greenhouse gas emissions: they claim the carbon dioxide is captured and permanently stored.

> **"Gas, like coal, has no future as the world wakes up to our climate emergency."**

The apparent marketing goal is to still use gas for energy used in homes and commercial buildings, but to convert the methane to hydrogen at remote facilities and then deliver hydrogen rather than gas through the pipeline system. The gas industry has regularly claimed that blue hydrogen is a low-emissions or zero-emissions fuel. If these claims were true, this could be viewed as a brilliant strategy for the gas industry, as it would allow them to keep their pipelines running and would entail a rather large increase in the sale of gas overall, since the conversion of methane to hydrogen is rather inefficient and requires more gas than is needed if the gas were instead to continue to be the fuel used in homes.[19]

Blue hydrogen is an illusion, another effort by the gas industry to distract and delay. I first heard of the idea of piping blue hydrogen to homes in a meeting in the spring of 2020 of the New York State Climate Action Council, a group on which I serve.[20] The Council is charged by law with implementing New York's progressive climate law. Emissions from energy used in our homes and commercial buildings, including both carbon dioxide and methane, are the single largest source of greenhouse gas emissions in New York State. Much of these emissions come from the use of gas for home and commercial heating in New York. The gas industry urged the Climate Action

Council to endorse "zero-emissions" blue hydrogen as a replacement for piping gas to homes and businesses for heating.

As it turns out, the industry claims on emissions were based totally on marketing, not on objective scientific analysis. Mark Jacobson of Stanford and I took this on as a research project and published the first peer-reviewed analysis of the greenhouse gas footprint of blue hydrogen in 2021.[21] Our conclusion? Because of less than perfect capture of carbon dioxide and because of methane emissions associated with the production and transport of the gas used to produce blue hydrogen, blue hydrogen emissions are far from near zero. Greenhouse gas emissions are, in fact, greater than if one simply used gas directly for heat energy instead. Blue hydrogen has no future in New York: the Climate Action Council, in our final blueprint plan for implementing the state's climate law, recommended that no hydrogen made from fossil fuels, with or without carbon capture, be produced or used in New York.[22]

> "We relied on real-world data on carbon capture and storage, while they argued that future technologies might improve the situation."

Predictably, the gas industry fought back furiously against our 2021 paper. Lobbyists worked to counter our message in the back channels of governments globally, with the press, and through social media. We were told our assumptions on both carbon capture and methane emissions were far too pessimistic. This is not true, as we used the best available data and performed sensitivity analyses across a wide range of possible assumptions: Our conclusions were hugely robust across all reasonable assumptions. And for many months, we were told that a rebuttal would be forthcoming soon in the peer-reviewed literature. That rebuttal was published in March of 2022 by Romano and colleagues, but the journal quickly retracted it since the authors had misled the journal about possible conflicts of interest: They had stated they had none, whereas several of the authors had deep and ongoing ties to the oil and gas industry.[23]

This critique by Romano and others was republished with corrected statements on conflicts of interest, and with our reply, in the summer of 2022.[24] The Romano et al. critique of our work had two major assertions: First, that we relied on real-world data on carbon capture and storage, while they argued that future technologies might improve the situation. And second, they argued our estimates for methane emissions were too high. This is categorically wrong and completely unsupported by any actual evidence they provided: The only citation they gave for methane emissions lower than those used in our analysis was from a cartoon figure on a gas-industry web site, with no links to any actual data. As Jacobson and I wrote in our reply, "There is no room for blue hydrogen in a decarbonized energy future. The Romano et al. result derives from a cartoon and hypothetical guesses, not scientific data."[25]

The gas industry continues to promote hydrogen as a major part of our energy future, often conflating the relatively low-emissions "green hydrogen" (made from electrolysis of water into hydrogen and oxygen, powered by 100 percent renewable electricity) with blue hydrogen, referring to both as "clean hydrogen."[26] Note that even for green hydrogen, the details of production matter hugely in determining the greenhouse gas footprint. In the United States, the Inflation Reduction Act of 2022 provides huge subsidies for clean hydrogen, with clean hydrogen defined as having emissions below four kilograms of CO_2 per kilogram of hydrogen produced.[27]

The gas industry clearly believes it can earn subsidies for blue hydrogen at this cutoff level, based in part on analyses such as those presented by Romano et al. in their critique of my paper with Jacobson, where using fanciful inputs they concluded emissions from blue hydrogen could be as low as 1.9 kilograms of CO_2 per kilogram of hydrogen produced. In comparison, our estimates based on the entire range of reasonable input parameters for carbon capture and methane emissions estimated blue hydrogen emissions as between thirteen and twenty-six kilograms of CO_2 per kilogram of hydrogen produced, with a best estimate of twenty. Clearly, the oil and gas industry has a lot to gain economically by lowballing greenhouse gas emissions. An executive for oil and gas giant Chevron recently stated (August 2023) that "gas is not a bridge fuel but a forever fuel,"[28] one that society will always use with carbon capture to make blue hydrogen fueled by massive government subsidies.

> "Several of the authors had deep and ongoing ties to the oil and gas industry."

What about green hydrogen? While many energy experts agree that it may play a role in a future decarbonized energy system, I and many others believe that the supposed centrality of hydrogen (including green) in our energy future is being overly hyped, driven at least in part by the public relations efforts of the gas industry.[29] While hydrogen can be used to store energy and can be used as a transportation fuel and even for heating, alternative technologies for all these exist and, in many cases, are likely to be more efficient and cost-effective. In the words of Professor David Cebon, a mechanical engineer at Cambridge, "the international fossil fuel industry, which is under existential threat from electrification, is promoting hydrogen as a solution to create confusion among politicians and the public and delay its own demise."[30]

Examining the use of hydrogen for home heating through delivery of the gas pipeline system, as pushed by the Hydrogen Council and large gas delivery utilities such as National Grid, is instructive. Simply put, this is a terrible idea for many reasons.[31] First, using renewable electricity to produce green hydrogen which then heats homes is very inefficient compared to using that electricity for heat pumps. For the same quantity of electricity, heat pumps deliver six to ten times more heat than does burning green hydrogen. As we move to decarbonize our energy systems, renewable electricity must be viewed as a precious resource, one to be used as efficiently as possible. Second, hydrogen makes gas pipelines more brittle, posing catastrophic risk.[32]

The California Air Resources Board advises against mixing more than 5 percent hydrogen with gas in a pipeline system because of this safety issue.[33] Third, hydrogen is the smallest molecule in the universe and, therefore, can easily leak from pipelines. Leakage rates of methane from pipelines can be quite high, and leakage of hydrogen would be greater. Although hydrogen itself is not a greenhouse gas, hydrogen leaked to the atmosphere does adversely affect the climate by interacting with greenhouse gases, for instance, increasing the residence time of atmospheric methane.[34] Clearly the best path forward, and the one recommended for New York State by the Climate Action Council, is to move rapidly away from gas and undertake an organized dismantling of the gas pipeline distribution.

The next few years are critical in how society tackles climate change. Will we embrace beneficial electrification and move rapidly away from fossil fuels? Or will we follow the false solution of continued reliance on the gas industry and their pipelines? We cannot leave this up to the fossil fuel industries.[35] Oil and gas have made their choice, and in the words of one of the officers of the Hydrogen Council and employee of the French gas company Air Liquide, "ultimately, blue hydrogen is an easier way for an oil company to pivot to clean energy than going full-on renewable. . . . It's a way to avoid having stranded assets from the current fossil fuel-based system."[36] The strategy of big oil and gas remains to distract, to delay, and to continue to rake in profits from their products that are destroying the world.

> "For the same quantity of electricity, heat pumps deliver six to ten times more heat than does burning green hydrogen."

Chapter Six

Opportunity Costs and Distractions

BF Nagy

Obviously, Professors Robert Howarth and Mark Jacobson and other experts are frustrated by government money being channelled into solutions that exist only in PowerPoint, created to delay real change. But they continue to patiently explain the science, economics, and failure statistics proving that carbon capture, "new" kinds of nuclear power, bioenergy, and geoengineering are distractions from real solutions. Investors are not fooled. They are walking away from these failed or unproven technologies toward wind, solar, and other renewables.[1]

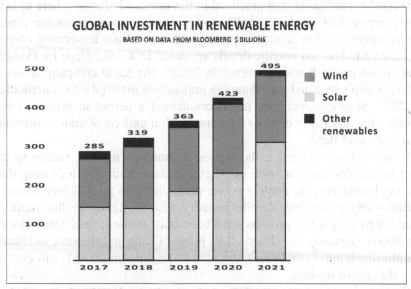

GLOBAL INVESTMENT IN RENEWABLE ENERGY
BASED ON DATA FROM BLOOMBERG $ BILLIONS

- Wind
- Solar
- Other renewables

2017: 285
2018: 319
2019: 363
2020: 423
2021: 495

Figure 6.1. Global Investment in Renewable Energy
Figure created by Climate Solution Group, using data from Bloomberg. https://about.bnef.com/blog/a-record-495-billion-invested-in-renewable-energy-in-2022/

POLLUTERS MUST PAY

Policymakers should remember the old adage: If you tax it, there will be less of it. We should enforce pollution laws, fine polluters, sue polluters,[2] regulate polluters, and, again, in the case of energy, consider nationalizing fossil fuel companies. Shift them to clean energy or wind them down.

In 2022, massive leaks of oil in Thailand, Peru, Ecuador, and Nigeria led to explosions, fatalities, fires, and extensive water pollution. The world's biggest tanker containing 1.1 million barrels of oil began leaking after being abandoned in the Red Sea near Yemen by a Chevron subsidiary.[3] Two years later, a project led by the United Nations finally drained the tanker and averted the threat.[4] Why couldn't oil companies have cleaned up their mess? Why won't gas companies accelerate the plugging of pipeline leaks, which pays for itself, and could significantly reduce emissions?[5] I recently happened on an article by investigative journalist Justin Nobel that uncovered the mostly untold story of careless stockpiles of toxic radioactive fracking waste in Texas. It's a stunning indictment of just how negligent the pursuit of profit is by these firms.[6]

NUCLEAR NIGHTMARE

Perverted capitalism is also practiced in the nuclear industry, where at least one company has decided to make money decommissioning nuclear plants, but apparently, it has limited expertise and a penchant for cutting corners to save cost. The full horrific details appeared in a solid piece by Douglas MacMillan in the *Washington Post* in 2022.[7] The same company wants to follow Japan's lead and start dumping radioactive material into waterbodies. In 2023, Massachusetts state regulators denied a permit modification that would allow discharge of more than one million gallons of toxic wastewater into Cape Cod Bay.[8]

In my opinion nuclear is the biggest technology mistake humanity has ever made. Yes, there is some beneficial nuclear medicine, let's keep that, but the bombs and the nuclear power plants have to go. Nuclear power is unbelievably expensive, despite industry disinformation to the contrary. That's why only a few projects have been built in the United States during the 2000s. Germany has closed all its plants. China and the state of Illinois are phasing it out. Even France, the world's leading proponent, will eventually shut down nuclear.

Sincere governments, investors, and energy professionals are tired of dishonest business cases followed by massive runaway cost overruns and decade-long construction delays. It's the costliest option for utility-scale power and takes the longest at fifteen-plus years to design and build, compared with about fifteen months for renewables. It doesn't work at all when the planet heats up. It's not safe, healthy, green, or clean.

NUCLEAR COST

If you were offered two identical cell phone plans, one for $180 each month and one for $50 each month, which would you choose? Nuclear power costs about $180 per megawatt/hour compared with $50 for wind and $60 for solar.[9]

Future Reliability—As I write this, half the nuclear plants in France are not operating because it's too hot and the water used to cool them is too warm. All electricity systems are vulnerable, and the most reliable are now renewables and batteries, supported by the kind of sophisticated load-management software described in chapter 12.

How safe, healthy, and clean is nuclear power? After seventy years, there remains no viable toxic-waste-disposal solution. It's poorly stored on nuclear plant sites near volatile reactors. That means thousands of years of continuing

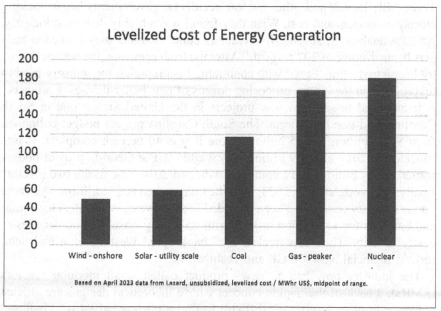

Figure 6.2. Levelized Cost of Energy Generation
Figure created by Climate Solution Group, using data from Lazard, April 2023. https://www.lazard.com/research-insights/2023-levelized-cost-of-energyplus/

costs to ratepayers to keep deadly nuclear waste material from releasing radiation that would kill us into nearby cities. These costs are going up because most sites are near populous places on seacoasts, where tsunamis, storms, and melting glaciers are growing threats. They are also vulnerable to terror groups and other bad actors, which are becoming increasingly aggressive and will eventually cause death and destruction.[10]

Deadly Plutonium-239 has a half-life of 24,400 years, an improbable nine-hundred-generation management assignment. Just two generations ago, in 1979, the United States built the Runit Dome in the Marshall Islands. Below an eighteen-inch concrete cap, they stored 111,000 cubic yards of radioactive debris from twelve years of nuclear tests. It's already cracking and leaking into the sea.[11] Meanwhile, back on US soil, for about forty years, the nuclear industry has failed to find a place to secure its waste. Nobody wants it buried under their town, and most municipalities don't want it transported through them. Most Millennials and Gen Z people don't want piles of nukes left behind when we all pass away, but it's already too late. There are about one-quarter-million metric tons worldwide, at least 80,000 metric tons in the United States. The latter costs more than $10 billion each year to safeguard. It's stored at more than seventy-five sites in thirty-five states.[12]

Recent Nuclear History—After death and disease caused by Chernobyl, Three Mile Island, and other nuclear accidents, governments began looking closely at dangers and cost. What they found was a highly dishonest industry. An EPA analysis of seventy-five reactors concluded that they exceeded budgets by an average of 207 percent.[13] After the truth came out, the nuclear industry began to decline. Faced with continuing bankruptcies, the industry tried to revive itself in the 2000s, proposing dozens of new better-than-ever projects.

It managed to sell two new projects in the United States: one in South Carolina, and one in Georgia. The South Carolina project budget ballooned from $11.5 billion to $25 billion before it was 40 percent complete. It was canceled in 2017 after $9 billion was wasted.[14] The Georgia project budget started at $14 billion with a completion date of 2016. One of the two reactors finally began operating in 2023, with the budget then at $31 billion. The same power capacity could have been provided in a few years with safe, clean solar and batteries for $17 billion or less, with no legacy waste and no decades of costly security. The "new nuclear era" bankrupted Westinghouse and dealt serious financial blows to GE and Toshiba.[15]

The industry now has a "new" product called small modular reactors (SMRs), a back-of-the-napkin concept whose theoretical designs are at least a decade from prototype. It's difficult to find a nuclear expert who believes in SMRs, apart from those who will benefit from them. "Delays, poor economics, and the increased availability of low-carbon alternatives at rapidly decreasing cost plague these technologies . . . there is no need to wait with bated breath for SMRs to be deployed," said Paris-based Mycle Schneider Consulting to *The Globe and Mail* in 2020, after it published its third review of SMRs in five years.[16]

HISTORY WILL BE WRITTEN

Other journalists and I will continue to tell the stories: naming the names, places, and details about those who are trying to save us, and those who would sell our planet to the highest bidder, waste our infrastructure dollars, sow destruction, and abandon millions of leaky oil wells across North America. Undoubtedly, these bad actors will try to re-write history the way authoritarian governments do, but every day, there are more honest storytellers joining us in both the free world and working under the radar in places where it's dangerous to reveal reality. These villains cannot escape their dastardly deeds. We won't let them.

ENVIRONMENTALISTS AND MEDIA:
WHAT WE GET WRONG

We environmentalists take the bait on disingenuous "debates" with profit-takers. We mindlessly adopt misleading terminology like "natural" gas, "alternative vehicles," or "radical protesters." We add to public confusion through complicated explanations full of obscure jargon. People just tune out. We seem to be unaware that we're losing an information war that matters.

Parents with young children, please watch the 2012 version of *The Lorax* by Dr. Seuss with your family. It's a fun movie that reminds me how brilliant was Theodor Seuss Geisel and how important it is to keep communication simple and interesting. By the time he died in 1991, Dr. Seuss had sold six hundred million copies of his books, which had been translated into more than twenty languages and had been enjoyed by billions. He wrote his most famous work, *The Cat in the Hat*, using Grade 1 vocabulary.

We who write about the climate emergency can no longer afford to be complicated, boring, negative, repetitive, and tangential. Let's stop talking about the end of the world using unintelligible charts; stop expecting people to care passionately about something that will happen in 2100. Our most important global warming goal (commonly discussed as 1.5 degrees or 2.0 degrees) refers to the change in average global temperature from pre-industrial levels in degrees Celsius. Snore. It's creatively and semantically flaccid, conjures up nothing in the human imagination, and tracks something that moves at a snail's pace. It's like telling me your library card number and expecting me to be engaged in the conversation.

We make it worse by slicing and dicing the details further, disagreeing with each other on miniscule points, and demanding front-page news for colorless events devoid of emotional content or story appeal. We're evidence-based, supposedly smart thinkers—but perhaps we are too smart for our own good; too intellectual to survive in today's media environment, too disconnected to keep things simple and artfully document the greatest developments in human history.

> **"Let's celebrate and illuminate the thousands of heroes struggling to reinvent all of our biggest industries, win against impossible odds and powerful villains."**

Let's celebrate and illuminate the thousands of heroes struggling to reinvent all of our biggest industries and win against impossible odds and powerful villains. My father once asked me why I would waste my time writing fiction. I tried to explain that stories synthesize the message better than reality does. Reality is messy. It's granular and confusing. Stories are simple, colorful, and memorable.

As mentioned, we're losing our ability to recognize non-credible information. We can't retain any idea for long before it gets bumped out of our conscious mind by something else. This is dangerous because the climate crisis imperative requires that we recognize genuine evidence from scientists and engineers and rapidly implement proven solutions. It starts with telling great human stories about how this is already being done.

Chapter Seven

Here Comes the Sun

Distributed Energy Resources and Virtual Power Plants

Dr. Audrey Lee, Laura Fedoruk, and Steve Wheat

If you're not familiar with virtual power plants (VPPs), you likely soon will be. They are popping up all over the United States and the world. A VPP is a group of homeowners who own solar panels, usually on rooftops, and modern smart home batteries, too. Aggregators and progressive electric utilities organize these homeowner groups and link the software in their dwellings to the advanced software used to manage the national grid, making it all work together, and optimizing efficiency. The model offers several benefits to ratepayers, electrical utilities, and to society as a whole. It cuts costs and emissions for everyone and reduces power blackouts. The three intelligent people who wrote the VPP section in this chapter helped invent the VPP in the United States while working together at Sunrun, an American solar supplier and aggregator.

With decades of unique experience, Dr. (Ja-Chin) **Audrey Lee** is a respected energy policymaker, economist, and executive, and she is among the foremost distributed energy resource experts in the United States. She led teams (with Laura Fedoruk and Steve Wheat) at Sunrun and at Advanced Microgrid Solutions that pioneered residential and commercial energy generation and storage systems and made them work with utility grids. She also worked as a Senior Economist developing policy with the White House, the International Energy Agency, and the California Public Utilities Commission.

Laura Fedoruk is now Technical Program Manager for Project Tapestry at X, The Moonshot Factory (formerly Google X).

Steve Wheat is Vice-President of Program Management at Swell Energy in San Francisco. More biographical information on Lee, Fedoruk, and Wheat can be found near the end of this book.

Humanity's slow response to climate change has put many of the systems to govern and manage our lives at a crossroads, including the electric grid. Change and adaptation has to account for many factors. These include lifestyles that are significantly more energy independent, technology that can be deployed in smaller and smarter iterations, and sufficiently sophisticated systems capable of handling inputs and outputs from millions of devices directly connected to an energy system designed more than a hundred years ago. In some cases, the change of the electric grid to a clean and resilient one will have profound impacts on the very survivability of huge swaths of the country suffering from one or many overlapping and highly dangerous climate-driven events.

One of the defining challenges for all of us, whether we have only been users of electricity, or have spent decades defining the laws that govern it, will be how quickly we can convert an aging energy system designed in times of relative calm to one that is more flexible, nimble, decentralized, and adaptable. This is especially true with a system that has become a patchwork of overlapping policies, regulations, and profit motives.

Distributed energy resources such as customer-owned solar and storage can play a primary role in the positive evolution of the electric grid. In order to balance increasing electrical energy use and limiting the impacts of climate change, the electrical grid needs to transform from a top-down, centralized system of large power plants to a more intelligent and networked system managing a two-way flow of information and energy.

To reach decarbonization goals and limit climate change, the grid needs to seamlessly integrate millions of fluctuating renewable resources and customer demand. It must evolve to a participatory partnership that increases resilience and societal benefits. Clean energy distributed energy resources (DERs) and the control, data collection, and services provided by advanced inverter technologies can help to integrate more renewables onto the grid while also improving the network.

> "The electric infrastructure relied on to power our lives is aging, failing, and producing dangerous emissions."

The electric infrastructure relied on to power our lives is aging, failing, and producing dangerous emissions. The impacts of climate change are already being felt on the grid around the United States and the world. Larger, faster-growing, and year-round wildfires have caused frequent grid outages, causing utilities to pre-emptively shut off power to millions of people in order to avoid the potential of sparking the next fire, such as the Public Safety Power Shutoffs (PSPSs) now practiced in California.[1] Even the electric grid's most reliable previous source of clean energy, hydropower from dams, has increasingly come under strain. Reservoirs feeding power plants, such as the Hyatt Power Plant, which was shut down in 2021 due to lack of water,[2] are sometimes falling below levels needed to generate power.

There is significant value in both electrifying and decentralizing our energy systems. The electrification of transport and buildings reduces direct energy consumption and greenhouse gas (GHG) emissions in those sectors and shifts them into the power sector,[3] increasing the impact of grid decarbonization.[4] The net effect is energy system-wide reductions in both energy consumption and emissions. By localizing energy production and control, problems on one part of our energy system can be prevented from cascading into the entire grid. Combustible fuel consumption decreases, protecting communities from the immediate health impacts, and communities who were previously only energy customers can become active partners in the creation of community resiliency assets.

Thankfully, as the threat of climate change has emerged and grown, the tools available at a personal, community, and national level have never been more advanced and more ready to tackle the challenge. DERs can be as simple as solar panels on a rooftop or a smart thermostat and as complicated as a virtual power plant consisting of tens of thousands of solar panels, batteries, and other devices across large areas that can be called on to fill the gaps as aging fossil fuel infrastructure is retired and an entirely new energy system of non-fossil fuels replaces it.

THE BENEFITS OF
DISTRIBUTED ENERGY RESOURCES (DERS)

Customers are making choices and investments today on a variety of DERs at their homes, commercial businesses, and industrial sites. They are increasingly pairing energy storage with solar.[5] DERs are sited and generate or manage electricity at or near the location where it is consumed, potentially providing locational grid value. This reduces the need for transmission and distribution infrastructure,[6] which is becoming increasingly difficult to site in urban and suburban places,[7] not to mention the physical losses associated with transporting electricity large distances from the place of generation to the place of consumption.

In this way, customer-owned DERs are inherently local and enable resiliency by avoiding outages related to equipment far from a customer. If a problem at one end of the grid occurs, energy can be created, stored, or delivered from somewhere else. And it provides the exciting potential to coordinate large numbers of DERs together (a "virtual power plant," or VPP) to react to challenges on the larger electricity grid and provide beneficial solutions to all customers.

Due to their localized nature, grid services provided by aggregated and coordinated DERs can result in faster emergency response than centralized power plants. It's like directly texting every phone in an emergency, rather than messaging a cable company.

The proliferation of DERs has upended the electrical grid model, initially built as a top-down, centralized architecture for delivering power from large power plants to small end-users. The ability to aggregate, monitor, and control decentralized assets is becoming critical to achieving the levels of renewable energy contribution necessary to avert catastrophic climate change while also cutting costs and maintaining grid resilience and reliability. Networks of decentralized energy resources can be monitored and controlled via advanced software systems to mimic larger-scale resources[8] despite being physically separate and even independently owned and operated.

> "Networks of decentralized energy resources can be monitored and controlled via advanced software systems to mimic larger-scale resources."

The evolution of the electrical grid from a centralized structure to one where energy can be supplied and served at the edges of the system from resources much smaller than power plants is a relatively new phenomenon, often referred to as the "Grid Edge."[9] This evolution has benefited from decades of work by planners, engineers, policymakers, and even university campuses—some of the first end-users to adopt microgrids.[10]

As the costs of technologies such as sensors, photovoltaic (PV) panels, and batteries have come down, their pace of adoption has surpassed even optimistic estimates. Reductions in infrastructure deployment costs such as solar PV have outpaced annual estimates by more than twice the estimated rate.[11] Significant and exponential growth has occurred in energy storage deployment across all market segments in the United States, and DER growth is also expected to follow these trends.

As an example of how rapidly technology adoption is occurring, between 2012 and 2022, net energy metering (NEM) connected solar in California experienced an average increase of approximately 30 percent per year.[12] In 2020, renewable energy, inclusive of rooftop generation and utility solar, accounted for over 33 percent of the state's total system generation.[13] In Australia, the uptake of rooftop solar has been even more dramatic, with one in four homes having rooftop generation installed.[14] Solar seems to be the driving force in displacing coal used in electricity generation in Australia.[15]

As the percentage of renewables on the grid increases, it becomes more necessary to consider policies to best coordinate distributed assets with the wider grid. There is a need for incentivizing the market and consumer behaviors that result in the lowest cost and highest reliability for customers across the entire grid, such as the pairing of intermittent renewables with firm assets such as energy storage. Generally, the lowest-cost marginal addition to energy grids and almost all of newly interconnected energy capacity is renewable generation. It is expected that distributed energy storage and PV capacity across Oceania, North America, and Europe will double between 2020 and 2025.[16]

For residential, commercial, or industrial customers, a battery combined with rooftop PV can provide electricity bill savings with time-of-use on-peak and off-peak electricity pricing arbitrage, as well as home electricity backup and islanding in the event of a grid outage. In addition, these customers can participate in a VPP program and more quickly recover the up-front costs of the battery.

Implementation of VPPs involves understanding the potential market mechanisms for participating DER assets of residential solar-plus-battery systems to receive payments for services, as well as analysis of options for optimization. For example, analyzing the group characteristics of a set of solar and battery storage systems, making use of their advanced metering telemetry and creating custom physics-based models. Analysis of the expected contributions of DERs combines engineering models with large empirical datasets from distributed systems and is key to unlocking the potential of these assets, while allowing regulators and industry to feel confident that these technologies can support the grid.

Incorporating analysis of historical time series DER meter data (customer electricity usage, PV solar generation, and battery storage charge and discharge) can lead to better operations of VPPs through better assumptions about asset availability and more awareness by grid operators of how these systems perform. It can help to find optimal ways to diversify a portfolio of assets included in an aggregation—such as patterns of energy consumption, and even the direction that solar PV panels are facing. Diversity of assets in a VPP, without the central failure points of large-scale power providers, enables a stronger and more resilient grid.

> "We need to accelerate project ideas and make them real."

We need to accelerate project ideas and make them real. Without detailed analyses, proposals are just ideas, and operational issues may never be uncovered and remedied. Models for forecasting and analysis will be critical to changing the operational nature of the grid, as will be finding ways to expand the energy markets to include unconventional assets. Unlocking the economic potential of these assets to provide multiple value and revenue streams—such as for both customer bill optimization as well as grid services of multiple types—will help to align incentives with reliability and lower carbon energy sources.

The adoption of solar energy by residential and commercial purchasers has been greatly enabled by the innovation of power purchase agreements (PPAs). This structure allows customers to access solar energy for a fixed cost per unit energy and transfers the cost burden of construction to a developer who benefits from a guaranteed price and market certainty.[17]

Going further, public policies that incentivize or subsidize DERs, and support customers in disadvantaged areas, have increased access to DERs at the same time as energy equity. These types of programs, like the California Self-Generation Incentive Program (SGIP),[18] California's Solar on Multifamily Affordable Housing (SOMAH3), and Multifamily Affordable Solar Housing (MASH4) can be essential to reaching these communities.

Policy, more generally, is crucial to creating the market structures that correctly value the benefits and services of coordinated DERs on the grid. This is especially true given the historical regulated monopoly model of the electric utility sector. Since regulated utilities profit from a fixed rate of return on capital expenditures, saving operational expenditure is not as attractive as capital outlay. So even if DERs and VPPs can decrease costs by deferring or replacing infrastructure upgrades—through reducing the need for new poles and wires via serving load near to consumption—they may not be attractive to utility bottom lines.

If utility profit structures disincentivize building local renewable energy and resilient systems, and if market structures do not allow for the profitable participation of non-traditional assets, there will have been a missed opportunity to enhance equity as well as progress towards climate change mitigation and adaptation. The Federal Energy Regulatory Commission's FERC Order No. 2222 enables several sources of distributed energy to be grouped together and collectively participate in wholesale energy markets, reducing market entry barriers for DERs.[19] It will hopefully help to increase the benefits of installed DER capacity as well as accelerate further deployment and new revenue streams. Policy change is an amazing accelerant because it can mean that assets that are already installed but being underutilized can quickly transform. Imagine if all the already connected grid assets such as home batteries, solar, EVs, and their advanced inverters were able to contribute grid services. The impact would be immediate as the infrastructure is already in place.

There is no stopping customer adoption of DERs, especially as costs decline and functionality improves. Both residential and grid-scale solar continue to reach new installation records.[20] But we also have an existing electricity grid into which all of us have already invested trillions of dollars. The optimal solution is to ensure that both operate together and complement each other, maximizing societal benefit.

The role of government and public policy is to create the rules, market structures, and incentives to achieve society's goals such as decarbonization, grid reliability, and customer access to electricity. This allows the private sector to innovate and thrive in these market structures. Stakeholder collaboration is key, such as having transmission and grid operators, states, and utilities work together to plan and connect more renewables to the grid while sharing the costs of doing so. These concepts build on policy successes in the distribution system and provide hope for the evolution of our shared grid.

This evolution is accelerating. As directed by the 2010 Assembly Bill 2514, in October 2013, the California PUC adopted a 1,325 MW procurement mandate for electricity storage by 2020. The procurement mandate was divided between three domains: transmission connected, distribution level, and customer-owned storage. Additional laws in 2016 increased the initial goal of 1,325 MW with a supplementary 500 MW target and helped the power providers reach this initial mandate, including through distributed energy storage system-specific mandates and incentives. The California Public Utilities Commission (CPUC) managed to approve procurement of projects to exceed the AB 2514 target and satisfy the domain-specific requirements.[21,22] As of mid-2023, the California Energy Commission (CEC) has reported that California connected energy storage increased by more than twenty times, from 250 megawatts in 2019 to five thousand megawatts in 2023, and connected capacity is expected to grow exponentially.[23]

Another policy win specific to customer-owned technologies, the Self-Generation Incentive Program (SGIP) began in 2001 through legislation to help address peak electricity problems in California. SGIP has provided incentives to a variety of distributed energy technologies, with eligibility and incentive levels changing over time to respond to California's evolving energy needs. A key part of unlocking VPP potential is the contribution of energy storage technologies to the program.

> "A key part of unlocking VPP potential is the contribution of energy storage technologies."

The CPUC has reported rapid cumulative growth of SGIP storage capacity and projects in both residential and commercial sectors in the years leading up to 2020. According to the 2019 SGIP Energy Storage Impact Evaluation, "Residential [battery] systems with on-site solar consistently provide benefits to customers in the form of billed energy savings during the summer, are discharging throughout investor-owned utilities (IOUs) and the California

Independent System Operator's (CAISO's) top hours and decrease GHG emissions, while utilizing only 60 percent of available capacity."[24]

Continued electrification provides acceleration to potential positive grid transformations. As homes and businesses are electrified with flexibly scheduled connected loads and DERs, such as solar, battery storage, and electric vehicles, the potential for VPP grid contributions will only increase.

Upcoming market reforms will have a positive impact. Disallowing exporting to the grid under market rules, a significant portion of the value could be stranded—i.e., a full battery may not be able to be used by the grid at all if net load is already at zero, because it cannot be "counted" and remunerated.

A 2018 analysis of stranded capacity was found to be significant—with both rooftop solar and battery storage potential to be unable to participate fully. That analysis did not take into account increased potential residential storage adoption based on possible grid services revenues or the ever-decreasing costs of technology. Both decreasing costs and increasing revenue potential will accelerate adoption and enable even larger potential for grid assistance via distributed solar and storage.

There is huge potential waiting on rooftops and in buildings that can make a difference to grid operation and planning, increasing resilience and decreasing grid carbon intensity. Customer-owned assets are uniquely suited to increase grid resilience because they are located exactly where they can be used if a grid failure occurs. These assets are a win for everyone—they can assist the grid when needed and provide customer resilience and customer cost savings. Including them and their potential grid services in comprehensive grid planning and operation makes sense.

HOW IS THE ROLE OF THE UTILITY CHANGING?

For many families across the United States, the home is considered their most important asset. It often appreciates measurably in value and offers a place to shelter and grow, engage in work and play, and fosters connections through schools, events, and a place of belonging.

A home has been the receiver of public goods and services. Homeowners pay taxes for public safety, education, sanitation, access, and, of course, electricity and gas connections. These one-way receiver relationships changed with the introduction of rooftop solar and additionally with the ability to store that solar in a battery capable of shifting energy usage throughout the day.

But even with a great deal of added renewable energy resources on the grid, and time-dependent utility rates that help to incentivize when home batteries can optimize load, in the end, each home acts randomly. Each solar system is set on a different roof in a different location, altered by the shadows of nearby trees, with batteries used either for backup power or bill savings. It represents a cacophony of energy, an unpredictable distribution system, and one with little visibility to the grid operator. It can become difficult to manage and optimize, for a utility whose goal is to deliver power twenty-four hours a day, seven days a week.

Each homeowner with a DER can shift their load patterns for the greater good, but without coordination, it would be difficult to systematically change the load pattern of even a local electric grid with so many unknowns. Initial solar generation on the distribution system is able to be instantly beneficial, but coordination and strategy are needed.

> **"Solar generation on the distribution system is instantly beneficial, but coordination and strategy are needed."**

Similarly, at the start of the sharing economy, millions of homeowners could see their in-law unit, or a spare bedroom, or their home itself in a completely different way. Parts of their home, or their entire home, could be put on marketplace platforms like Airbnb to generate income and help increase the number of rooms available for visitors to come and patronize local businesses and tourist spots.

Everyone with a car, many Americans' second most valuable asset, could download an app and their vehicle, previously parked and unused for 90 percent of the time, could be activated as part of a vast network of vehicles. This could allow everyone to participate in improved energy asset efficiency with the simple click of a button.

The sharing economy has caused great change, and the community grid, virtual power plants, and aggregations of DERs have a chance to be permanently beneficial—taking the changes to our energy system to an entirely different level. It is a solution that scales and allows benefits to multiply. If enough assets are included in a VPP, it is possible to rival fossil-fueled power plants, and even avoid the need for new transmission and distribution lines.

As an initial benefit, the aggregation of DERs represents a tool to help with acute, short-term issues, like everyone turning on their air conditioners on a hot day and straining the grid or risking a blackout. Yet they also become an important tool for the entire energy system and its planned operation, the sum much greater than the accumulation of their parts.

There are two important initiatives that lay the foundation for today's diverse, clean energy grid. The first is to create new standards for data and transmission, giving visibility and predictability to grid areas that have been traditionally unknown. The second is the development of a smarter grid by today's progressive utility operators.

To make DER and smart device data accessible and actionable will require involvement of everyone from asset manufacturers to utilities, system operators, government agencies, and non-governmental organizations. IEEE 2030.5 and open-source projects through LF Energy (Linux Foundation) stand out for creating data transmission standards that protect consumer privacy while accommodating grid efficiencies and analytics. When all of our DERs and connected devices speak the same language as the utilities that are charged with managing our electricity, then the second important piece, smart grid, can fall into place.

The standard name for the organizations that manage the collective use and operation of DERs is "aggregators," and they are the groups whose job is to manage thousands of connected DERs to operate in concert with each other and meet the energy or grid needs of the community around them. Aggregators are most often private energy companies, and homeowners who enroll in VPPs are paid for allowing the aggregator to change the operations of their devices for this purpose. The assets added to their homes and businesses that were installed in order to save on monthly utility bills now begin generating additional value by making their communities more resilient.

This benefit doesn't solely apply to those that can afford homes and the installation of advanced energy systems. Anyone who lives in an apartment building with EV chargers installed and who owns an electric vehicle can become part of the energy community. Anyone with a smart thermostat in their apartment can also make important contributions. And the benefits of a neighbor's system can accrue to everyone.

When an aggregator gathers devices together, it can share the information about how they operate with the utility so that the cacophony of energy moving back and forth seemingly at random can become a symphony, harmonized with every other tool on the grid. These might include hydroelectric dams, wind turbines, nuclear generators, and vast fields of utility-scale solar projects and utility-scale batteries.

The benefits for all of us sharing in this energy system continue to grow as aggregators collect more DERs. The energy services team at Sunrun completed a study[25] that found that as few as seventy-five thousand solar and battery homes, in the Los Angeles region, which serves more than 1.4 million customers, networked together could replace approximately three hundred megawatts of peaking capacity of one of LA's retiring gas plants.

"This benefit doesn't solely apply to those that can afford homes."

The blistering, planet-wide heat wave in the month of July 2023 became an ever-widening arc of worry about the climate. Research by Katherine Hayhoe at Yale found that while 70 percent of people were concerned, 50 percent felt helpless and didn't know what to do about it. The ability of individuals to contribute to proven climate solutions is one of the great benefits of DER aggregation at scale. Every person involved can identify their meaningful contribution. Imagine the hope engendered if the thousands of people in a virtual power plant received a message that because of their efforts a fossil fuel power plant was retired and re-wilded into greenspace.

PUTTING THE SMART IN SMART GRID

The promise of VPPs that are large enough, smart enough, flexible enough, and robust enough to replace huge fossil fuel power plants on the grid is revolutionary.

VPPs involve customers and power utilities and also require an aggregator, whose role largely relates to information technology and data management, because getting from opportunity recognition to implementing VPPs is still a daunting task.

With every device type, model, manufacturer, as well as the countless varieties of customer-associated utility bill structures, and massive troves of data being created by each device, new capabilities need to be developed to synchronize them all into a virtual power plant. A well-coordinated and optimized grid can account for home energy demand and usage, real-time solar conditions, weather variability both local and regional, grid voltage conditions, frequency, current, operational states, and charging and discharging schedules.

How the industry collects, uses, stores, and analyzes this data, whether to convert it to something operable in utility planning and forecasting, optimizing for customer savings (in the case of solar and batteries), customer comfort (in the case of smart thermostats), or device utility (in the case of EV chargers), all has to be balanced and managed individually and in aggregate, creating formidable challenges and exciting opportunities.

As the number of DERs sold and installed multiplies, the use cases for how they can continue providing more and better tools for managing the energy system rise. As this potential increases, the incentive conflicts between very different groups may as well.

How can we continue to make the devices inexpensive enough to reach larger groups? How can we balance differing agendas such as backup power, utility bill savings, range anxiety, and home cooling? How do we ensure that policymakers, regulators, and utilities can confidently rely on and manage a brand-new kind of grid? And how can we juggle these competing concerns, overlapping consequences of market dynamics, policy, and regulatory changes that often take very long amounts of time against the urgency of building out VPPs as quickly as possible?

Smart data analysis helps with numerous objectives related to grid optimization, carbon reduction, emergency backup power, energy cost savings, investor asset performance, and potential new market inclusion mechanisms. Thinking smartly about what data to use, store, and surface for different purposes becomes part of asset management and the creation of a VPP, along with privacy and security.

A step to answering many VPP operational questions is the creation of a software tool, a distributed energy resource management system (DERMS). This system can be developed in-house by an aggregator or purchased off the shelf from a software as a service provider, but importantly, it is a tool that can communicate directly from an aggregator to a utility, and vice versa, to help manage individual devices and entire fleets of DERs.[26] These tools often represent a platform that is capable of including wide varieties of DERs and can be adapted not just to communicate directly with utilities but also with energy wholesale markets that can overlap dozens of states and utilities simultaneously (ISOs).

> **"Smart data analysis helps with numerous objectives related to grid optimization."**

Even with the tools to effectively aggregate and manage all of these devices, the amount of data remains enormous. The electric grid has to be managed in real time, twenty-four hours a day, seven days a week, so effectively collecting all of this data, analyzing it, making sense of it, and making split-second decisions for managing the grid remains a provocative challenge for everyone in the business of managing DERs.

As our team at Sunrun began building larger and larger fleets of DERs, the data volume and complexity continued to grow. Across every part of the company, from financing solar and battery systems, designing the right combination of hardware at each home, accounting for geography, climate, utility rates, and historical patterns of already built systems, it became clear that an entire team of data scientists and new data infrastructure was necessary to collect and synthesize the data, as well as make sense of how to use it to improve investor confidence, operations, and customer experience, in addition to the VPP operations themselves.

One example of these operations working successfully at Sunrun was the National Grid ConnectedSolutions Program. The program started as one that paid incentives to residential customers for using their smart thermostats for demand response. We created a program that offered customer payments to discharge their batteries up to a set number of times during summer to meet the energy demands on the hottest days, when everyone usually turned on their air conditioners.

This program benefited from a straightforward design and relatively simple utility rates for the customers who enrolled. This meant the concept of a VPP could easily be explained to customers, even in a market where batteries were a brand-new product. The utility benefited by knowing that stored energy would be used at the time when it was most needed, and each year, VPP participants were paid by check for their contributions.

After the program's first year was completed, surveys showed that customer opinions of their utility improved as a result of seeing their systems used for more advanced functions. The program continues to grow and remains a success story in the industry and country.

THE INFLATION REDUCTION ACT
AND OTHER KEY INITIATIVES

The electrical grid is the largest and most complex machine ever devised—one that must instantaneously balance supply and demand every moment of the year. According to the web site of Southern California Edison (SCE), California's second-largest utility, "SCE maintains more than 105,773 miles of distribution lines. Our system contains approximately 1.4 million electricity poles. If you were to lay down the wires that makeup SCE's extensive transmission and distribution network end to end, they would traverse the United States approximately 40 times."

The system is also old, dating back to Thomas Edison's Pearl Street station in Manhattan in 1882. This lumbering behemoth doesn't just have to keep up with the changes posed by transitioning to renewable energy, it also has to maintain its operations in hundreds of overlapping marketplaces, with different rules, laws, and policies at the federal, state, and local level. It needs to decentralize itself so that power plants and large facilities once thought to be built in "safe" areas are not threatened. It faces new formidable threats such as floods, wildfires, and occasional superstorms, plus traditional threats such as squirrels and vegetation. The challenge of modernizing this grid is overdue, huge, and complex.

> "The program continues to grow, and remains a success story in the industry."

But the tools to replace these technologies are already here, already active, and deployed in the field. The responses needed to scale and connect them will require action at every level in order to employ new forms of energy generation, storage, and grid-balancing power electronics.

The Inflation Reduction Act (IRA 2022) and the Infrastructure Investment and Jobs Act (IIJA 2022) are bringing about huge advancements as the largest climate investments in history. And great progress has been made and must continue to be made by federal agencies.

For example, the EPA must continue driving cost accountability of fossil fuel externalities and cleanup. The Federal Energy Regulatory Committee (FERC) must continue building on the national rules for integrating DERs into energy marketplaces, such as FERC Order 2222. The Department of Energy Loan office must continue to bridge the gap between the DER technology capabilities we have today and the bankability of full market acceptance of DERs as VPPs in the near future so that aggregators and others in the space can accelerate their efforts at reining in the complexity of aggregation management. ISOs must continue refining their own rules and regulations, speeding up connections and reducing the barriers to entry for smaller developers and aggregators to enter their marketplaces alongside larger projects.

At the state level, policy is also needed, harmonizing requirements across fifty state public utility commissions. At the local level, individual utilities, whether investor-owned, municipal, co-ops, or community choice aggregators, need to invest in their own DERMS capabilities to integrate DERs, work with aggregators to solve localized energy system issues with all the tools DERs offer, and help create frameworks to properly compensate them for that participation.

Aggregators must unite all of these threads at the customer level and to help grid planners. They must become more advanced, sophisticated, and flexible. They must communicate well with customers whose devices are participating, utilities and ISOs relying on these devices as a tool to manage the grid, and the policy stakeholders at the local, state, and federal level.

Unfortunately, in many instances, a patchwork of DER companies has emerged, each building a vertical stack of data, monitoring, and control software,[27] rather than working more collaboratively. We must work as an industry to create a wide platform ecosystem to seamlessly connect disparate DERs.

It will take all solutions, from top down and bottom up, to truly scale operations quickly enough to meet the demands of a changing climate. There is no height at which a walled garden will be preserved if the entire energy system is not changed.

Even without net metering incentive programs, residential rooftop solar and battery storage are economically attractive in places like Hawaii. As electricity rates continue to increase due to infrastructure upgrade expenditures, the technology becomes more economically competitive.

But these technologies are not being integrated into the grid to their highest potential. This is in part due to a policy failure, which disallows their deployment in markets where ill-equipped operators fear their use threatens effective control of the grid. Focusing on resilience and reliability will be imperative—and there is potential here for DERs to improve the grid in these aspects. The proliferation of renewable energy technology is both a mitigation and adaptation strategy, and adjusting the incentives of grid participation can help to scale beneficial solutions.

Evolving the grid into a two-way, renewable energy infrastructure that is equitable, resilient, and cost-effective is a highly interdisciplinary problem with many stakeholders and many opportunities for high-value input.

Learning to use DERs to their fullest potential is critical to efficient evolution of the grid. Rapid progress has been made in the DER space. Between August 2020's rolling blackouts, when some DERs operated under a state-issued request to assist grid stability,[28] and the summer of 2022 extreme heat wave in California, distributed battery deployment nearly doubled. Those distributed assets were then ready and able to help support the grid in a meaningful way—with output of approximately a mid-sized power plant being deployed during critical peak hours.[29]

"The biggest battery in the world is . . . located in garages around California."

The executive director of the California Solar and Storage Association was quoted as saying, "The biggest battery in the world is . . . located in garages around California and they are helping keep the lights on for everyone."

The IRA contains key measures that will help residential, commercial, low- and moderate-income, affordable housing, environmental justice, and tribal customers adopt DERs, directly through tax credits, loans, grants, rebates, and loan guarantees and indirectly by supporting manufacturing of DERs.[30] The potential capabilities of DERs keep growing and will be continuously added. We need to keep up the momentum, to scale at the pace needed. And our individual impact must contribute to the success of others on the grid, to which we are all connected.[31]

Chapter Eight

Change

BF Nagy

This book is filled with solid evidence—references to academic and case studies, economic and climate trendlines. But sometimes we don't need to verify what we know instinctively is right. Audrey Lee grew up in Los Angeles, with unforgettable smog days in the summer. She now lives in San Francisco with a giant Monterey Cypress outside her window. "The fog condenses on the cypress leaves and drips into the ground, yet the tree struggles to continue to coexist with the non-porous pavement and houses we've built," says Lee.

"My maternal grandmother died when I was in high school, and my mother organized a road trip for our family to get away and unwind after all the stress of caretaking and mourning. I have fond memories of going to Sequoia National Park and just taking the time to breathe that clean, amazing air in the middle of the majestic trees. I could see the positive effect of the forest and environment on my mother, especially, since she had carried the burden of caring for her mother through lung cancer . . .

"I think humans have evolved to value things that are scarce, but our sense of scarcity is misleading, as our impact as a society has expanded beyond what we see and feel immediately around us. I'm interested in creating and improving energy systems, markets, technologies that are more efficient, and can sustain society into the future." Lee has two sisters, one daughter, and one son and says, "The imperative for us is to act now for future generations."

"Growing up on Vancouver Island on a farm with horses, and near the water, shaped my sense of connection to nature," says Laura Fedoruk. "Some of my best memories are of long days of chores coupled with rides on the beach and forestry roads . . . ringing a bell to let the horses know it was dinner time and feeling the thundering of hooves on the ground as they raced up from pastures at the bottom of the hill.

"My recent experience of living in California and seeing the devastating toll of wildfires, and when in September of 2020 there was a day the sun never appeared to rise in the San Francisco Bay Area. . . . And now having two young children, I feel the need to work on resilient, efficient, and renewable energy systems. I can't imagine working on anything else, and my sense of urgency to help halt the climate crisis is the strongest it's been."

In 2023, Steve Wheat told me: "I spent my twenties teaching English as a second language abroad and chasing the ghosts of a world that disappeared as I fell in love with it. When I lived in Japan, I taught Hemingway's *The Snows of Kilimanjaro* the first year in modern history that the snows had completely disappeared. In Shanghai, we lived between rivers that ran through the city the color of antifreeze.

"It became abundantly clear that the amount of work needed to preserve what was special was overwhelming, and I spent my time after I returned to the US trying to find my place in those efforts. Our health and well-being as well as our economic opportunity and movement towards a sustainable human civilization are all interdependent, and all require us to get completely off of fossil fuels as soon as possible."

Wheat is also a talented poet who has published pieces in several magazines, including a haunting poem about New York, flooded by very deep water.[1] Not only are these young people motivated to make change, but they also have the skills and intellectual horsepower to make an impact quickly. We need only to get out of their way and let them get on with the job.

Change is uneven. Some people are still living in the 1900s, while others routinely use technologies that will not become mainstream for another decade. Most of the people running this world are too old. Before you accuse me of ageism, please note that I don't care what your birth certificate says. To me, an "old" person is someone so risk averse and resistant to change that they should retire now from any position of influence in business or government.

People who don't like change are often also afraid of their computers and phones and like to stick to what is working well enough to get them through another day. They're perfect victims for those who would spread disinformation because they might find it difficult to verify something they've read online. Before buying an electric car, electric heat pump, solar panels, or voting for the "slightly radical" green person for whom they should vote, they prefer to "wait and see." Unfortunately, what they're seeing while waiting is the planet collapsing around us. Let's end fossil fuels now. Let's end fossil fuel subsidies now. Let's stop talking about false solutions. We have real proven solutions that work.

TRANSPORTATION FOR TOMORROW

"I took up fly fishing in the mid-1990s," says former California Public Utilities Commissioner Dr. Nancy Ryan. "I especially love the Carson River in California. Also the Walker." She told me that when she first started fishing in mountain rivers, she was surprised to discover the impact of hydropower development on trout habitat and the viability of their principal food source. In the Sierra Nevada mountain range, much of the main rivers' flows are diverted through pipes and flumes, dropped through turbines, and then diverted again. Only a relative trickle remains in the natural riverbed.

Her response revealed her inner economist: "I developed models to demonstrate the (minimal) cost of replacing electric generation that would be lost if more water was allowed to stay in the rivers to nourish trout. This work was the beginning of my thirty-year career as an environmental advocate, pro-environment energy regulator, and consultant supporting clean energy and transportation."

The East Walker River is described by an online blogger as "one of the best fly-fishing streams in California . . . home to plenty of good pocketwater and chubby trout that reward an angler's patience." Not far from Lake Tahoe, the beautiful Carson River begins its journey in the towering Sierra Nevada in California, winding gently into Nevada. Near Hangman's Bridge, there are two hot spring pools, in case you need some outdoor rest and relaxation.

"I like to be tuned out of society and tuned into nature when I go fishing," says Ryan. She says she would prefer to substitute renewables for hydropower when possible. "It doesn't seem right to me to sacrifice the life in our rivers."

MY TOUGH-GUY JOURNALIST FRIEND, MICHAEL

Not typical compared with many environmentalists, Michael Barnard's approach is to optimize climate solutions that also yield economic and business opportunities, strip away wishful thinking, and develop strategic scenarios for the long term. He had a military upbringing, advises multi-millionaires on where to invest, and comes off as gruff, savvy, and wickedly intelligent.

The first crack I noticed in his armor came when he told me that while working in the financial industry in his early twenties, he happened upon the Brundtland Report from the World Commission on Environment and Development, and it made a big impact on his thinking. He was already becoming concerned about the planet: "I wanted to make a difference to the future of the world, not a difference to a bank's bottom line."

He also admitted to being conflicted between his support for the North American Free Trade Agreement (NAFTA), which could build the economy, and sympathy for those who opposed the lumber operations at Clayoquot Sound in the province of British Columbia (BC), one of the most beautiful old-growth forests on the face of the Earth. A trip to British Columbia in the mid-1990s cemented his concerns: "I was driving to Tofino and came across a clear cut after wandering through Cathedral Grove. I walked out into the clear cut and cried."

ELECTRIFYING TRANSPORTATION

Electric vehicles of all kinds are already far better quality products than conventional combustion vehicles that come with compelling economics and deep decarbonization. And global sales trends show we are rapidly electrifying personal transportation. We've passed adoption tipping points in most countries and demand is streaking skyward on an aggressive S curve.[2] The electric vehicle (EV) story is proving the heroism of the majority of humans. It's a narrative about smart government policy, but also more about bold moves by ordinary people. As mentioned, today's most inspired economists know that when transforming a market economy, consumer-driven demand is more powerful than government supply-side intervention.

Whatever we may think of CEO Elon Musk and Tesla, they deserve a historic nod for doing far more, against impossible odds, for commercialization and technical proof-of-concept of EVs than anyone else. Everyone has and continues to follow Tesla.[3] It led the effort to create global demand for electric vehicles, reinvented car manufacturing and marketing, and is leading innovation in other transportation modalities.

During early adoption, government incentives can be powerful catalysts for change, along with herd mentality. EVs and solar panels represent key elements in changing our culture because they are so visible. Studies show that when someone puts solar panels on their roof, some of the nearest neighbors soon follow.[4] The same thing happens with EV adoption.

These two are thus among the highest priorities for government in the next ten years because they can lead the dramatic shift of our global mindset. They are visible, create collective spirit and peer pressure, make big impacts against emissions, and offer the opportunity for ordinary people to contribute to and demonstrate their support for change. If we keep the electrification shift going, adoption of the other clean technologies should follow effortlessly. As the younger generations increase their power, resistance to the deployment of climate solutions will dissipate.

Our current priority should be to electrify cars and other vehicles; however, we are also redesigning cities to reduce the number and size of cars in use. Personal vehicles will likely continue to be popular in the countryside, but pilot tests during the pandemic showed that the best cities are decreasing vehicle use downtown.

Let's reclaim the streets for people and minimize parking lots. Quality of life improves quickly in urban environments that shift priorities to public transit, pedestrians, cyclists, and others using more efficient vehicles than cars. Cars are incredibly wasteful in terms of energy, space, and return on investment. It's easy to envision a future where an AI-coordinated fleet of autonomously operated and efficiently shared cars and shuttles belong to the municipality's transit or freight network, performing last-mile or first-mile delivery of people and goods between homes, businesses, and electrified rapid transit nodes.

There are four immediate key challenges for municipal transportation planners. One, catalyze rapid electrification of public and private transportation. Two, increase investment in public transportation. Three, tackle affordable housing shortages by significantly reducing exclusionary zoning, which causes car-priority urban sprawl. Four, redesign downtowns for better quality of life while overcoming resistance among drivers who have spent years racing through towns in cars and trucks.

A top initiative for electrification is to help eliminate a key barrier to EV adoption, a perceived lack of public charging. The perception matters even if the reality is that around 90 to 95 percent of charging will take place at home and business locations.[5] The National Renewable Energy Laboratory (NREL) is planning on the basis of thirty to forty-two million EVs on American roads by 2030, requiring 182,000 fast charge ports and one million slower public ports.[6] As of 2023, there were about two million EVs on US roads. In 2022, about ten thousand fast charge ports and fifty thousand slow charge ports were added.[7]

These numbers suggest that although network development is trending faster, it's still too slow. The most experienced EV strategists, like Dr. Nancy Ryan, former California Public Utilities Commissioner, are looking further ahead. Dr. Ryan recognizes the twin goals of transportation electrification and fewer conventional private vehicles in cities and proposes policy planning for optimized efficiencies through EV grid integration (please see chapter 9).

Governments must also manage transportation in harmony with other transitioning systems, create regulatory frameworks for driverless vehicles and numerous other emerging types of personal vehicles, drones, and robots.[8] They need to educate planners, managers, and inspectors on clean energy and new planning tools equipped with artificial intelligence, redesign urban parking approaches, and repurpose downtown parking lots. They can help create urban beltway distribution hubs for efficient downtown food and goods delivery, help manage food waste and circular economy programs, help develop effective enforcement regimes for construction, land and water transportation, and support modern ways among developers and transportation professionals.

Chapter Nine

Policy and Planning for Electric Vehicle Grid Integration

Nancy E. Ryan, PhD

Dr. **Nancy E. Ryan** of eMobility Advisors is an economist with more than thirty years experience in the electricity industry. She is focused on the low-carbon transformation, with emphasis on transportation electrification and renewable energy. She served as Commissioner for the California Public Utilities Commission, Partner at Energy and Environmental Economics in San Francisco, and taught economics at UC Berkeley's Goldman School of Public Policy. She holds a PhD in Economics from UC Berkeley and a Bachelor of Economics from Yale.

The automotive and electric power industries are converging. Born in the late 1800s, they traveled separate roads and grew to vast scale by the close of the twentieth century. Utilities bought cars and trucks to support their operations, and the auto industry purchased electricity to light offices and power factories. Today, drivers increasingly see electric utilities as the enabler of reliable and affordable mobility. In the twenty-first century, transportation electrification will drive electric utilities' load growth, as the rapid adoption of air-conditioning did in the mid-twentieth century.

Mass electrification of transportation presents both challenges and opportunities for electric utilities, energy regulators, electric vehicle (EV) drivers, and the host of companies that offer innovative charging solutions. Integration of EVs onto the electric grid has the potential to deliver economic and environmental benefits to both EV drivers and the general body of utility customers. Notably, EV charging is a highly flexible load that has the potential to help balance intermittent wind and solar generation, lowering the cost of decarbonizing electricity generation.

What does this win-win outcome look like? For EV drivers, success means no-compromise mobility,[1] enabled by convenient, reliable, and affordable charging coupled with opportunities to save on charging costs by providing valuable grid services. For utility customers and energy regulators, success means that utilities serve EV charging load in a cost-effective manner that delivers economic and environmental benefits to *all* utility customers, supports utilities' efforts to modernize their systems and decarbonize their resource mix, and takes advantage of technological advances in the latest generation of EVs to enhance grid reliability and resilience. For innovators, success means public policies that create opportunities for them to profit by delivering technologies and services that appeal to drivers and enhance the grid's efficiency.

To reach this future, we must successfully navigate the intersection of EVs and the grid. The next section outlines the benefits and costs of EV adoption—from the perspective of EV drivers, electricity customers, and society at large. It then turns to two key aspects of the vehicle-grid nexus, implementing rates and managed charging programs for EVs and advancing the technology and standards that enable maximum flexibility benefits. In addition to policy recommendations, it highlights recent research and regulatory actions and identifies resources on lessons learned from pilots, including best practices.

COSTS AND BENEFITS OF LIGHT-DUTY EV ADOPTION

Electrifying light-duty vehicles can save money for drivers and utility customers while reducing overall emissions. Over an EV's lifetime, savings from avoided gasoline purchases and lower maintenance costs more than offset higher electricity bills and its up-front price premium. Spreading the fixed costs of electricity generation and distribution infrastructure over increased electricity sales yields surplus revenue that can be used to lower rates or invest in the grid. Substituting domestically produced electricity for imported petroleum-based fuels reduces our exposure to volatile global oil markets, brings energy jobs home to the United States, and stimulates local economies. As our electricity supply becomes cleaner, electrification also reduces net life-cycle emissions from the transportation sector. A growing body of research demonstrates that consumers are already realizing these benefits and that they will only grow over time.

> "Numerous benefit-cost studies have found that switching from a gasoline-powered vehicle to an EV reduces drivers' total cost of ownership."

The value proposition for EV drivers is favorable in most parts of the United States today, and it is steadily improving. Numerous benefit-cost studies have found that switching from a gasoline-powered vehicle to an EV reduces drivers' total cost of ownership (TCO).[2] The savings are greatest where electricity is relatively inexpensive but persist even in states with high electricity prices.[3] Where EV rates and managed charging programs are available, drivers can save even more by charging off-peak or providing grid services.

Currently, state and federal purchase incentives help offset the higher upfront prices for EVs; however, EVs are becoming more affordable as a wider variety of models become available and battery pack prices fall. Accounting for about one-third of an EV's price, battery packs are the principal driver of EV purchase costs. Analysts generally agree that EVs will reach purchase price parity with conventional vehicles when the price of battery packs reaches $100/kWh.[4] Between 2010 and 2021, battery pack prices plunged from $1,000/kWh to $132/kWh. Prior to a recent uptick driven by supply chain disruptions and higher raw material prices, most analysts expected that ongoing advances in battery chemistry and cell design would further propel this trend and prices would hit the $100/kWh by the middle of this decade.[5] Although reaching this milestone may take a few years, it is important to note that many of the forces exerting upward pressure on battery pack prices are also contributing to higher gasoline prices and increased sticker prices for conventional vehicles.[6] Also, elevated prices for critical metals are stimulating investment in mining and alternative battery chemistries.

Widespread EV adoption can also benefit a utility's overall customer base, if the right rates, programs, and policies are implemented. Revenue from charging creates headroom in rates, provided that the cost of serving that load stays below the average cost of service. Surplus revenue from EV charging can be passed along to the utility's customers through lower rates or invested in grid modernization to enhance efficiency, reliability, and resiliency. It can also help pay for chargers or grid upgrades to support EV charging. The key to capturing these savings is providing incentives for EV drivers to charge outside of peak hours, when renewable energy is abundant, or at times when the local distribution system is not congested.

Forward-looking benefit-cost studies have consistently demonstrated the potential for utility customers to benefit from EV adoption, and they have shown how effective EV rates and managed charging programs can boost those savings. M. J. Bradley & Associates (now ERM) weighed the costs and benefits of EV adoption in nineteen states. They consistently found that the net present value (NPV) of net benefits (NB) of EV adoption was positive and that it was higher when some drivers were assumed to participate in EV rates or managed charging programs. EV adoption yields benefits to all utility customers because modeled revenue from EV charging exceeded the estimated cost of serving the load (including any utility investment in charging or supporting infrastructure).[7]

Retrospective studies confirm that the projected savings are indeed materializing. Melissa Whited et al. used reported utility data to compare electricity rates paid by EV drivers to the marginal cost of providing the energy plus expenditures associated with utility EV programs. Nationwide, during 2011–2020, total revenues from EV charging exceeded total costs by $1.7 billion (in 2020 dollars). This finding held across regions, although the lion's share of savings was realized in California, home to the country's largest EV population.[8]

Driving on electricity is almost always cleaner than driving on gasoline or diesel fuel. Full battery-electric and fuel-cell vehicles produce no tailpipe emissions, and plug-in hybrid EVs emit far less greenhouse gases and air pollutants than conventional vehicles. They also eliminate or reduce indirect emissions from crude oil production and refining.[9]

On the other side of the ledger are carbon and conventional air pollution produced from electricity generation. Where and when a driver charges their EV determines the carbon content of the electricity that powers it. This is because the resource mix varies widely across the United States, and the marginal source of generation typically varies hourly and seasonally. Researchers have found that even taking these emissions into account, an EV is typically responsible for lower levels of greenhouse gases (GHGs) and other pollutants than an average new gasoline car.[10] And although manufacturers continue to increase the fuel efficiency of conventional vehicles, the carbon emitted from burning a gallon of gasoline or diesel fuel cannot be reduced. In contrast, policy and market forces are steadily reducing the carbon intensity of our electricity supply.

> "The inherent flexibility of EV charging makes EVs an increasingly valuable resource for grid operators."

Combined with the greater efficiency of electric motors, the trend toward cleaner sources of electricity—away from coal and toward natural gas and renewable power—is expanding the gap between tailpipe and smokestack emissions. EVs can also smooth and lower the cost of this transition. The inherent flexibility of EV charging makes EVs an increasingly valuable resource for grid operators. Historically, summer and winter peak loads, driven by air-conditioning and heating, respectively, drove electricity resource planning decisions. Today, even as temperature extremes grow and fluctuate more widely, grid operators must also build flexibility into the grid to offset the variability of the wind and solar generators that are increasingly replacing fossil-fired power plants. Grid operators need flexible resources to balance the supply and demand of electricity in real time as output from wind and solar facilities varies on scales that range from seconds to seasons. Flexible resources include quick-starting natural gas combustion turbines, hydroelectric plants with upstream storage, large stationary batteries, and EVs. For example, provided they have access to workplace charging (or work from home), many EV drivers can charge during midday hours when solar power is abundant and avoid charging in early evenings when solar generation drops off. On a shorter timescale, starting and stopping charging, or even discharging an EV's battery to the grid, are ways that EVs can help balance minute-to-minute fluctuations in wind and solar generation.

Finally, researchers have found that the "emissions debt" associated with an EV battery is typically "paid off" in the first few years that the vehicle is driven.[11] These emissions mainly come from the energy used in battery manufacturing, mining and production of critical materials, and end-of-life battery processing. They are also declining over time as domestic battery manufacturing ramps up. This is because our generation mix is much cleaner than in China, the main source of most EV batteries today. Similarly, onshoring mining and processing of critical materials is reducing the embedded emissions in EV batteries.

Rates and managed charging are the key to capturing the benefits of EVs— EV rates and driver-centered managed charging programs are the key to realizing the potential economic and environmental benefits of EV charging. Well-designed residential tariffs and programs deliver cost-based incentives to use the grid efficiently, share savings with participating EV drivers, and do not compromise their mobility. Although not discussed here, getting commercial and industrial (C&I) rates right is also essential to attract investment to electrify fleets and build out the public charging network.[12]

Residential EV charging is a large load, often using as much energy as the driver's entire residence. It is also flexible and smart. EV drivers only need to charge their car for a few hours per day, and they tend to charge at home if they can.[13] Telematics, the systems of sensors, microprocessors, and communications capabilities commonly available on today's vehicles, provide an inexpensive means to manage charging and measure an EV's electricity use.

To maximize the economic and environmental benefits of electrifying personal transportation, utilities have to get three things right about EV rates and managed charging programs: availability, awareness, and accessibility. Utilities should make *available* a variety of *optional* EV rates and programs. Effective marketing, education, and outreach are crucial to ensure that EV drivers are *aware* of these opportunities and appreciate the potential bill savings. And, EV drivers need inexpensive, hassle-free ways to *access* the cost savings afforded by these rates and programs.

*EV rates and managed charging programs should be available to all drivers—*All EV drivers should have the opportunity to leverage their charging flexibility to provide grid services and receive compensation. Flexibility can be monetized via targeted EV rates or demand response (DR) programs, participation can be enabled with a variety of technologies, and control can be implemented by the customer directly or remotely by the utility, OEM, or a third party.

Time-of-use (TOU) rates, with reduced cents/kWh charges during off-peak hours, are a sensible point of departure. Easy to understand and follow, they introduce drivers to the concept of saving money by shifting charging, and they deliver a threshold level of benefits to the grid. Pilots have consistently shown that EV drivers respond to TOU rates, especially when they have access to enabling technology.[14] Technology can enable either "passive" or "active" charge management; in the former, the customer schedules charging sessions in response to a price signal, perhaps using the timer built into their vehicle or wall charger; in the latter, a third party shifts charging, usually subject to constraints and parameters set by the driver.

> **"Pilots have consistently shown that EV drivers respond to time-of-use rates."**

Smartphone apps provided by OEMs, utilities, and third parties increasingly offer drivers active charge management, making it easy for them to maximize bill savings without risking the possibility of an incomplete charge. Active charge management also helps minimize the risk of "timer peaks," which can occur when a population of EVs simultaneously start charging at the beginning of the off-peak or super off-peak period.

Whether the rate or program applies only to the EV or to the driver's entire home is a critical distinction. Separating out the EVs' energy use on drivers' utility bills spotlights the savings that come from shifting charging. Since the public's perception of EVs and drivers' charging behavior are still in the formative stages, highlighting these benefits will be crucial in the coming years. Available evidence suggests that TOU rates that apply only to the vehicle (EV-only) are more effective in shifting load than whole-house TOU rates.[15] Fortunately, approximately half of the fifty-four residential EV TOU rates identified in a recent survey applied exclusively to the EV. Unfortunately, all required participating customers to have a costly dedicated metering device for their EVs.[16]

Over time, utilities can introduce more complex options that yield additional grid benefits and offer higher compensation for participating EV drivers. More sophisticated time-varying rates and demand response (DR) programs create value by shifting charging in response to day-to-day (and even hour-to-hour) fluctuations in electricity market and grid conditions.[17] These rates and programs are usually implemented via remote "active" charging.

Utilities have piloted a variety of optional rates, including day-ahead, locational real-time pricing, peak-time rebates, and event-based DR programs.[18] Drivers participating in DR programs usually have the ability to opt out of events. These more complex rates and programs are not for everyone; they should be open to all EV drivers, but participation should be voluntary, and drivers should be allowed to opt out of events. Also, some EV drivers may be able to provide multiple types of grid services that are mutually exclusive. Allowing them to "stack" revenues enhances the value proposition for EV ownership and augments savings in the utility's cost of service.

Finally, promptly transitioning successful pilots to mass-market programs is crucial to achieving cost-effective vehicle grid integration as EV adoption accelerates. In a survey of EV drivers, most reported that they would be willing to participate in EV rates and/or managed charging programs.[19] Yet, only about 25 percent of residential customers had access to time-varying EV rates as of 2019, and about a third of the available rates were still pilot programs. Most residential managed charging programs are classified as pilots, with enrollment capped at a few hundred or thousand EV drivers.[20]

New York is an important exception: In 2020, the New York Public Service Commission ordered the state's utilities to scale up successful pilots and subsequently issued a decision that approved their proposed mass-market programs and directed them to explore options to separately meter EV-charging load.[21] Elsewhere, the majority of EV drivers have little or no opportunity to receive compensation for shifting charging to low-cost hours, and the general body of ratepayers is missing out on the associated cost savings. Also lost is the opportunity to shape consumers' expectations and behavior around charging while EV adoption is still at an early stage.

Effective marketing programs are needed to make drivers aware of opportunities—EV drivers need to know that there are attractive opportunities to save money on charging in order to benefit from them. Yet, even in states with the highest EV adoption, awareness is low. For example, a survey of Northern California EV owners found that only 24 percent of EV drivers and just 30 percent of drivers with networked Level 2 chargers were aware of DR programs or of their benefits.[22] Utilities can improve awareness by enlisting dealers and automakers in marketing EV rates and programs to their shared customers.

EV drivers also need low-cost, hassle-free ways to access EV-only rates and DR programs—To participate in an EV-only tariff or DR program, drivers have to be able to separate their charging load from their overall electricity use. Until very recently, the only authorized way to isolate EV-charging load was by installing and, in some cases, purchasing a second revenue-grade meter dedicated to the vehicle. This is both costly and inconvenient, deterring drivers from opting into EV-only rates.[23] Yet, about half of the EV rates identified in the Lawrence Berkeley National Laboratory (LBNL) survey required drivers to have such a device. There are lower-cost and more convenient ways to isolate EV-charging load.

Advancing key technologies and standards can enhance EV value to the grid—Policymakers should capitalize on two key innovations that are rapidly transforming the ways EVs interact with the grid. Telematics, the integrated system of sensing, processing, and communicating capabilities common in today's vehicles, enables remote charge management and delivers high-quality data that could be used for billing purposes. EVs with bidirectional charging, such as the Ford F-150 Lightning pickup truck, can discharge electricity to the grid, power a home during an outage, or provide clean, quiet electricity at a campground or job site.

"Policymakers should capitalize on two key innovations that are rapidly transforming the ways EVs interact with the grid."

Fully commercializing these technologies requires programs, policies, and standards that allow customers to monetize the enhanced grid services they can provide and support a sound business case for OEMs to invest in developing them and installing them on vehicles. Technical standards also ensure the safety of consumers and utility personnel, protect customers' privacy, and support cybersecurity efforts.

VEHICLE TELEMATICS

Today, most new light-duty vehicles are equipped with a host of sensors and microprocessors, Wi-Fi, and, in some cases, GPS trackers. Telematics-based managed charging programs use vehicle data to optimize the timing of charging, enhancing the value of an EV as a grid resource. Like stationary chargers, telematics-based systems can monitor the rate of charge (kW) and total energy dispensed (kWh) in a charging session. Unlike stationary chargers, they have access to the vehicle's location and its battery's state of charge *at any point in time*.

By factoring state-of-charge into the optimization algorithm, telematics-based managed charging systems can determine how long it will take to replenish the battery and, therefore, how much latitude there is to turn charging off and on in order to shift load and provide grid services. Telematics data also provide insights into individual drivers' charging habits and travel patterns, which help customize charge management and enable optimization of charging across locations.

Utilities across the United States have piloted a variety of managed charging programs that use telematics data and systems. In California, BMW and PG&E's ChargeForward pilot demonstrated using telematics to schedule participating vehicles' charging sessions to support grid reliability and prioritize charging with renewable energy.[24] Figure 9.1 shows how data from the vehicle can be used to develop a charge plan: drivers specify their charging preferences (e.g., least cost, lowest emissions, etc.) and planned departure time in a smartphone app, and the system schedules charging.

Telematics data can also be used to measure an EV's electricity use, eliminating the need for a costly second utility meter. Already several utilities, including National Grid,[25] ConEd,[26] and DTE,[27] are using telematics-based systems to implement incentive programs to encourage drivers to charge outside of peak hours and verify eligibility for incentive payments. The next step is to develop the rules, standards, and policies to use more granular telematics data to calculate bills for drivers enrolled on time-varying EV rates. EVs' battery management systems already measure to a high degree of accuracy when and at what rate and for how long electricity flows into the battery; data transmitted to OEM servers via onboard Wi-Fi can be adjusted as needed and then passed on to utility billing systems.

Certifying telematics systems as mobile submeters would lower the cost of implementing EV rates and DR programs and increase their enrollment numbers. Mobile submetering has no incremental cost because it takes advantage of equipment that is already built into the vehicle. It expands the pool potential of participants because it enables a "bring your own charger"

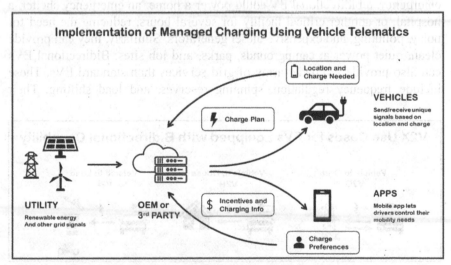

Figure 9.1. Implementation of Managed Charging Using Vehicle Telematics
Figure created by Nancy Ryan.

approach: Drivers can use whatever type of charger they prefer and would not need to purchase and install a second utility meter or networked Level 2 charger just to enroll. This more inclusive approach makes it possible for renters, multi-family residents, and low-/moderate-income EV drivers to save money on charging without incurring any incremental cost.[28] Mobile submetering complements stationary submetering, as some customers prefer to purchase a networked Level 2 charger.[29] Utilities and regulators in California, New York, North Carolina, and a handful of other states are beginning to explore the viability of using telematics as mobile submeters.

BIDIRECTIONAL CHARGING

Bidirectional charging: vehicle-to-grid, vehicle-to-home, and vehicle-to-load—Many of the new EV models arriving over the next few years will be capable of bidirectional charging, commonly referred to as V2X. With the ability to discharge their battery directly to the grid (V2G), a building (V2H), or equipment (V2L), they can provide many types of services to the grid, the public, and drivers. See Figure 9.2. The range, scale, and value of the services bidirectionally equipped EVs can deliver will only grow as battery storage capacity continues to increase.

Bidirectional EVs are essentially mobile storage devices that can provide the same services as a stationary battery. An additional advantage is that they can travel to sites where electricity is most needed. In an outage or other emergency, a bidirectional EV could power a home, an emergency shelter, a hospital, or another critical facility for several hours, reducing the need for noisy, polluting, and expensive diesel generators. Similarly, they can provide clean, quiet power at campgrounds, parks, and job sites. Bidirectional EVs can also provide a broader array of grid services than standard EVs. These include frequency regulation, spinning reserves, and load shifting. Their

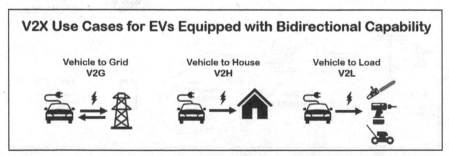

Figure 9.2. Examples of V2X Use Cases for EVs Equipped with Bidirectional Capability
Figure created by Nancy Ryan.

ability to charge and discharge multiple times while plugged in significantly augments their value relative to standard EVs. Putting appropriate incentives in place will enhance V2X's value proposition, driving adoption of bidirectionally equipped EVs.

Safety is a paramount consideration whenever a distributed energy resource injects energy into the grid or powers a building or piece of equipment. Interconnection processes and associated technical standards are needed to protect utility personnel, customers, and the general public. Fortunately, it is not necessary to start from square one: V2X systems can usually fit within existing interconnection processes. For example, islanded systems (V2H, V2L) can fit into processes for backup generators, and V2G systems can fit into processes for exporting stationary energy storage systems.[30]

With the right policies, programs, and tariffs in place, electrification of cars and light trucks can save money for EV drivers and utility customers. Getting there will require utilities and regulators to scale up rates and programs that have been successfully piloted, with states that have lagged behind skipping the pilot phase. Benefits will increase as utilities and regulators put in place standards that fully exploit the potential of telematics and bidirectional charging technologies to grow participation and increase the range and value of grid services.

Chapter Ten

The Energy Storage Challenge

Professor Jeff Dahn and BF Nagy

Professor **Jeff Dahn** of Dalhousie University is recognized as one of the pioneering developers of the lithium-ion battery that is now used worldwide in laptop computers, cell phones, cars, and many other mobile devices. From the time that he started doing research to the present, Dahn and his teams have worked closely with Tesla, NOVONIX, and other progressive firms. He is a Professor in the Department of Physics and Atmospheric Science and the Department of Chemistry at Dalhousie University and is invited as a keynote speaker to virtually all significant battery-technology conferences. He has won numerous prestigious awards. Details on these and his other career highlights can be found near the end of this book.

Storage, for electricity and thermal energy, is the most important problem facing humanity today. Not because we don't know how to do it; on the contrary, recent advances are astonishing. The challenge for human beings is to dramatically increase adoption of clean, renewable energy, partly by ensuring that storage technologies are available when needed, and that they are as efficient, as capable, as safe, and as inexpensive as possible. These are some of the goals at the lab at Dalhousie as part of a global science and engineering community that continues to improve solutions and optimize each kind of storage for its ideal applications.

Dalhousie has been among the research and development leaders in the advanced battery field for many years. The research part, into physics fundamentals, has involved numerous collaborators on more than seven hundred papers on lithium-ion batteries and other kinds of storage. The development part involves prominent battery users, primarily Tesla. The group has a long record of taking research beyond the lab and working with industry on technical engineering aspects to develop and refine practical solutions for real-world conditions.

For example, in addition to working with Tesla, a local company, NOVO-NIX (Professor Jeff Dahn is Chief Scientific Advisor of NOVONIX), has teamed up with Emera Technologies in Halifax and the US Department of Energy on a microgrid platform called BlockEnergy.com that will provide seamless interfaces between utilities and microgrids that can include rooftop solar plus supplemental ground-mounted solar and batteries. The platform will initially be deployed in several locations in the southern United States.

Positive electrode materials that were invented in the lab at Dalhousie are still used in some lithium-ion batteries today. This lab has been part of the development community that made lithium-ion batteries an emerging force in electricity storage. The Nobel Prize in 2019 for lithium-ion demonstrates that its advances have made a significant impact on human society. Lithium-ion batteries are integral to the utility of mobile phones, laptop computers, cordless power tools, electric vehicles, and, increasingly, all other main transportation modes and grid energy storage.

The "Million-Mile" Battery—The Dalhousie lab's goal for advanced batteries is decades of useful life. The "million-mile battery," "four-million-mile battery," and "hundred-year battery" are popular media concepts, and they are also scientifically possible with lithium-ion batteries.[1] A million-mile battery would last at least three times as long as an average combustion vehicle. The Dalhousie team and others have proven that they are possible with successful lab experiments and commercial products with appropriate chemistry that exists today.

> "The 'million-mile battery' and 'hundred-year battery' are popular media concepts, and they are also scientifically possible with lithium-ion batteries."

Battery costs continue to drop significantly, and electric vehicle (EV) companies are patenting processes and offering many options for longer range, faster charging, and longer battery life. Some seem to be already offering batteries that last much longer than the other parts of the vehicles in which they are used.

Dalhousie researchers work at eliminating the failure of lithium-ion cells due to parasitic reactions between the charged electrode materials and the electrolyte that slowly consumes electrolytes and electrode materials. Due to these parasitic reactions, the amount of charge stored in the battery during the recharge, known as the recharge capacity, is very slightly greater than that delivered during the discharge.

Additives to the electrolytes of lithium-ion cells and coatings on the surfaces of electrode particles can now dramatically extend useful life and cycle life of the cells. The way that additives improve cycle and calendar lifetime is not understood in detail, in spite of the publication of hundreds of journal papers from authors around the world. The Dalhousie team couples fundamental studies of additives and their impact with advanced diagnostic methods to move rapidly forward. The lab's precision coulombic efficiency measurements allow the rapid ranking of the effectiveness of electrolyte additives and surface coatings.

The team published a paper in 2019 that looks at ways to eliminate unwanted parasitic reactions between the electrolyte and the charged electrode materials that slowly, over time, degrade the materials. It is also attempting to detect and quantify parasitic reactions and make them smaller through changes to cell chemistry.

In one test, single-crystal graphite cells[2] began running in October 2017 at room temperature and continued into 2023, realizing at least 5.5 years of continuous cycling with only about 5 percent degradation at 1°C. These results can also be achieved at a higher temperature of 40°C with seven years of continuous cycling and 30 percent degradation over that time. The factors contributing to the cells' long lifetime include switching from polycrystalline to single crystal, the choice of quality artificial graphite, and appropriate electrolyte additives.

By 2022, the team demonstrated the impact of upper cutoff voltage on cell lifetime.[3] At the International Battery Seminar in Florida, the team showed that NMC811 cells greatly outperformed the Dalhousie LFP cells, which, at that time, were seen as equivalent to the best commercial LFP cells anywhere. If a lithium-ion cell can operate for more than six months at 85°C, how long can it last at ambient temperature?[4]

SODIUM-ION BATTERIES

Because it can take a long time to commercialize each kind of battery and manufacture them in large numbers, we must be thinking ahead about materials, chemistries, and processes. At the present time, lithium-ion batteries are dominating the electric car and electric truck markets, and they may continue to do so for a decade or more; however, beyond that, alternatives are likely to be needed. Adoption of EVs is increasing extremely quickly, and lithium, although abundant, is not available everywhere, and is therefore relatively expensive, depending on where it is found and where it is used.

By 2030, the world could require four hundred terawatt-hours (TWh) of storage, but it may only be producing six terawatt-hours of lithium-ion batteries each year. If the lifetime of a battery is only ten years, humanity will have to manufacture forty terawatt-hours each year, beginning immediately, which is a formidable challenge. Benchmark Mineral Intelligence estimates $200 billion to build factories to make just six TWh. Still, for context, US military spending is $700 billion per year. It's all about priorities. The imperative is to move on from fossil fuels or the cost will be much higher than these numbers. The needed forty TWh per year of battery production for grid energy storage decreases quickly if batteries can be made that last longer, which is why something like a million-mile battery is so important. Ultimately, when four hundred TWh of batteries are deployed, there will be a need for five hundred million tons of positive electrode material and five hundred million tons of negative material in that fleet of batteries.

There are now an increasing number of battery material possibilities for light ground vehicles, including sodium-ion batteries, solid-state batteries, and others. At the moment, sodium-ion batteries seem to be a strong option, with some manufacturers already introducing them in a few small-car models.[5] It will be necessary to dramatically increase the use of this or a similar material instead of lithium, nickel, and cobalt. Sodium is both abundant and available all around the world and thus should remain relatively inexpensive. Fortunately, we have more than thirty years of learning with lithium-ion, and it is similar enough to be useful in the development of sodium-ion.

> "At the moment, sodium-ion batteries seem to be a strong option."

So far, sodium-ion batteries have been shown to have different strengths and weaknesses when compared to lithium-ion batteries. They have lower energy density than lithium-ion and provide shorter driving range but should have comparable low-temperature performance and cycle life compared to lithium-ion. The newest sodium-ion batteries do not require scarce materials like cobalt and only a small amount of nickel.

Low energy density is currently a challenge for sodium-ion batteries; however, they will likely improve rapidly, similar to the way lithium iron phosphate (LFP) batteries were deficient in their energy-storage capability initially, and have quickly become a suitable option for electric vehicles.

Sodium-ion batteries may become a good option for lower-priced, high-volume passenger vehicles, especially in cooler climate regions. At the Shanghai auto show in 2023, Contemporary Amperex Technology Co. Limited (CATL) announced that sodium-ion batteries would be installed in the Chery iCAR, while BYD said a sodium-ion battery would be in mass production in the second half of the year for use in its Seagull models. Initially, these were powered by LFP batteries.

It's important to completely transition away from fossil fuels. The world needs electrified transportation and grid storage for energy from solar panels and wind turbines, for use when the wind is not there and the sun is not shining. The batteries for grid applications can be larger; however, they still need to last as long as possible, be as inexpensive as possible, and be as sustainable as possible.

Chapter Eleven
Aviation Electrification[1]
Michael Barnard

Michael Barnard is on the Advisory Board of electric aviation firm FLIMAX, has worked with several other aviation startups, and provides future-oriented decarbonization technology and investment guidance to multi-billion-dollar industrial firms, infrastructure funds, and venture capitalists. His assessments of aviation and maritime technology reform, global grid storage, and decade-by-decade projections of demand and solutions through 2100 have gained significant attention in multiple industries. He publishes regularly in CleanTechnica.com and Illuminum .com, and his work has also appeared in *Newsweek*, *Forbes*, and *New Atlas* and has been published in several textbooks and nonfiction books about decarbonization.

Aviation is a major source of greenhouse gasses (GHGs), punching above its weight due to additional forcing factors of contrail- and nitrous-oxide-related warming, in addition to direct carbon dioxide emissions. It's in the same range of total greenhouse gas emissions as hydrogen manufacturing, as a hard-to-decarbonize segment. It's going to decarbonize, whether the industry likes it or not, because the climate implications are sufficiently significant that it must.

But how it's going to decarbonize is a different matter. There are many contenders to be the aviation fuel of the future, and I've assessed most of them. Hydrogen would double the cost per passenger or freight mile and introduce very significant safety concerns as -249° Celsius cryochilled hydrogen will have to be inside the air frame with passengers. The combination makes it infeasible, in my opinion.

There are other equally intractable problems with aluminum air fuel cells, another contender being touted by one of the refueling startups. Being nonrechargeable and requiring the depleted batteries be returned to an aluminum smelter to be remanufactured means heavy weights of aluminum will be traveling further than just the air travel, and aluminum smelters are not evenly spread at all. Once again, infeasible.

The contenders for replacement which pass basic tests of viability are sustainable aviation biofuels and the increasingly energy-dense cell batteries that are already fit for four-hundred-kilometer trips. Some SAF biofuel processes are supplemented with hydrogen, and that's where the demand line for hydrogen fits in.

Late in 2021, I published version 1.0 of my projection of aviation refueling through 2100 based on analyses, discussions with aerospace startups and engineers, and assessment of the impacts of multiple strong trends. Subsequently, I engaged in discussions with more experts and analysts to consider how to refine the next iterations.

The most substantive discussion was with Wilma Suen, PhD in strategic alliances and, most recently, the global Vice President of Portfolio Strategy & Forecasting for General Electric's (GE's) airplane-leasing business. Until very recently, it was the largest such organization in the world. She and her time-zone-distributed team spent their days analyzing the best data that a multibillion-dollar organization could buy, to project flights per person per year for every country in the world for five, ten, and twenty years. Her insights led me to a slight increase in growth.

There are several things to call out in this projection. First, aviation is unlikely to return to pre-COVID-19 levels quickly. The efficiency and effec-

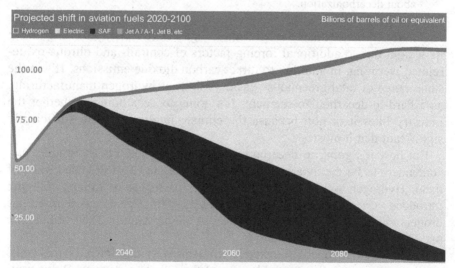

Figure 11.1. Projected shift in aviation fuels, 2020–2100
Courtesy of Michael Barnard.

tiveness of videoconferencing and remote work has been proven across the economy, but especially in those sectors that used to be heavy fliers, and I say that as someone who used to fly weekly for business. We've moved a decade into the future and broken the habit of getting on planes, and corporations will no longer pay for rafts of consultants to show up Monday and fly away Thursday. Client-facing now means over video links for the vast majority of the information, sales, and consulting workforces.

Among many other data points, this was reinforced by a recent discussion with a former colleague who now runs the national ESG practice for Canada for one of the major consultancies. While their junior staff are itching to get back to client sites, that's just not going to happen. Only the most important and critical deals will involve flying from now on, and then, only a subset of them.

Similarly, in-country travel is going to be much more favored than international travel by many more people for vacations, and that will persist. COVID-19 was the fourth epidemic in the past twenty years, following SARS, H1N1, and Ebola, and now everyone knows the consequences. COVID-19 wasn't the last epidemic; it was just the worst epidemic since the Spanish Flu.

Both IATA and Boeing project in the range of 4 percent growth year over year for the next twenty years. I consider this deeply unlikely after the post-COVID-19 bounce. Population growth is slowing and will end between 2070 and 2100, depending on which demographic projections you prefer, with the United Nations being on the conservative side. That means fewer new potential passengers.

Alternatives to flying are increasing as well, specifically high-speed rail, with China alone having built forty thousand kilometers of electrified passenger service since 2007, California's high-speed rail having started construction, and Morocco planning to expand its current 323-kilometer service to 1,500 kilometers in the coming years. Shorter-distance travel will be subsumed somewhat by autonomous electric ground transportation as well, as people sleep their way to their destinations. Finally, aviation refueling will come with increased costs through the second half of the century, putting a strong economic inhibitor on increases in demand. However, global affluence continues to grow, and with that, new customers.

The combination means that I've added a slow rise per decade to demand through 2090, where it flattens out, instead of being flat after the post-COVID-19 bounce. This is a global perspective, with significant geographic variance.

Second, electric means battery-electric but is not limited to lithium-ion chemistries. Lithium-ion is fit for purpose for aircraft with turboprop engines which carry up to nineteen passengers as far as four hundred kilometers. Density, cost, and weight equations will continue to improve every year. The fast-charging technology for cars and trucks is fit for purpose for rapid recharging of small electric airplanes today.

The amount of manufacturing, research, and innovation underway in battery technology means that lithium-ion is improving roughly 28 percent for every doubling of manufacturing capacity. But new chemistries are coming that will see significant improvements. Lithium-ion will likely prove suitable for up to one-hundred-passenger planes for one thousand kilometers, but other chemistries will take over.

> "Lithium-ion is fit for purpose for up to nineteen-passenger turboprops with four-hundred-kilometer ranges. . . . Density, cost, and weight will continue to improve every year."

Only in the second half of the century will a mixture of aviation advances occur, including greater battery density, better power management, novel electric engines currently on the drawing board, and improved airframe aerodynamics that will enable cross-Pacific aviation.

Third, battery-electric and biofuels are only going to start making a noticeable dent in aviation fuel toward the middle of the 2030s. This is a hard sector to decarbonize, and significant transformation and development is required to achieve these targets. For example, in my piece articulating the case for biofuels, one of the transformations that will occur is the elimination of wasteful use of biofuels in easier-to-electrify sectors. The two million barrels a day currently going into cars and trucks needs to be repurposed as all ground transportation is electrified.

The economic intersection of battery energy density, weight, and cost is sufficient for four-hundred- to six-hundred-mile-range planes with four to nineteen passengers today, and that will grow substantially with each passing decade, but it's insufficient for the majority of current flights right now. And, of course, battery-electric passenger planes need to be manufactured and certified, on which companies like Heart Aerospace and ELECTRON Aviation are working.

> "Hydrogen is going to be providing very little energy for aviation. I see no future for it as a directly used fuel."

Fourth, I've put a line in for it, but sharp eyes will note that hydrogen is going to be providing very little energy for aviation. I see no future for it as a directly used fuel based on the assessment of massive increases of cost and the need to use hydrogen for much more useful decarbonization purposes. However, hydrogen is used in some biofuel processes to supplement the hydrogen bound up in the cellulose, and there will be a window for some economically viable use of green hydrogen for that purpose.

Fifth, sustainable aviation fuel (SAF) will only grow as a source through 2060, when I project that energy densities of batteries will be sufficient for 80 percent of flights and a sufficient number of battery-electric airplanes will be replacing the current fleets. After 2070, it starts diminishing rapidly as fuel-burning turboprops and turbofans retire. Bio-fuels for aviation are only a forty-year growth market, and after that, they will diminish. Of course, agricultural lobbies will work to extend this, and they may very well pervert or at least defer rational decision-making.

> "Energy densities of batteries will be sufficient for 80 percent of flights."

Of course, many analysts have issues with this perspective. People seemingly close to current SAF results have disagreed with my earlier analysis. Others deeper into batteries have disagreed with my perspectives on aluminum air fuel cells, albeit on what appear to be technicalities and some lack of a systems engineering perspective. People invested in hydrogen for their various reasons disagree with this analysis, of course. It is one of many potential scenarios and necessarily imperfect.

Greenhouse gas emissions—What is important with this analysis are the implications for greenhouse gas emissions (GHGs). Similarly to the projection of fuel usage, I'll call out a set of obvious observations and some not so obvious ones.

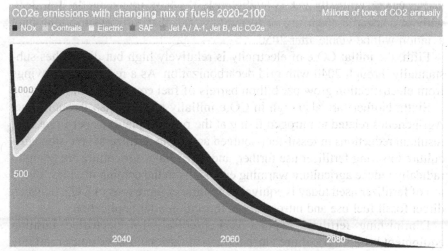

Figure 11.2. CO_2e emissions with changing mix of fuels, 2020–2100
Courtesy of Michael Barnard.

First, aviation-related global warming emissions plummeted during COVID-19. These include direct carbon (CO_2), nitrogen oxides (NO_x), and contrail clouds (about one-third of aviation emissions impact) because enormous numbers of flights didn't occur for several years. It's relatively insignificant compared to the history of CO_2-equivalent (CO_2e) emissions that will drive warming for the next century regardless of what else we do, but it's still there.

Second, while aviation energy requirements will not return to 2019 levels until mid-century, it will still rebound a fair amount over the next eight years, and emissions will rise as a result.

Third, it will likely take three decades before net emissions for aviation are back below 2020 levels. That's a problem, but they'll still be well below 2019 levels. That said, this projection suggests that while aviation is a hard target, more attention needs to be paid to it. Aviation emissions rising substantially this decade is not a climate solution.

Fourth, contrails and NO_x CO_2e ratios make it clear that biofuels are not a long-term solution. Both are significant warming side effects of burning fuels. The model assumes operational changes that substantially reduce contrail formation and impacts, something that can be substantially done by just five-hundred-meter changes in operating altitude. But burning any fuel in our atmosphere produces NO_x with 265 times the global warming potential of CO_2. Biofuels substantially reduce CO_2 from the fuel and operational changes reduce CO_2e from contrails, but NO_x persists. They are much better than fossil fuels, however. That said, electric drive trains would emit no NO_x and create no contrails and so is obviously the best-case scenario. I project, without a specific chemistry or technology set, that fully electric long-haul aviation will be viable after 2050.

Fifth, the initial CO_2e of electricity is relatively high but diminishes substantially through 2040 with grid decarbonization. As a result, CO_2e savings from electrification grow per billion barrels of fuel equivalent with time.

Sixth, biofuels are also high in CO_2e initially but diminish through time. Agrigenetics related to nitrogen fixing at the roots of plants as per Pivot Bio, resultant reductions in fossil-fuel-sourced ammonia fertilizers, precision agriculture lowering fertilizer use further, and low-tillage agriculture are going to radically reduce agriculture warming emissions in the coming decades. Every ton of fertilizer used today is equivalent to nine or more tons of CO_2e between direct fossil fuel use and nitrous oxide formation after application.

Diminishing fertilizer applied very precisely by electrified farming equipment brings agriculture close to zero emissions. I have conservatively not assumed soil carbon sequestration from farming in this analysis, but many studies project the potential for agriculture of all types to become carbon negative. This wouldn't change the need to eliminate biofuels due to contrails and NO_x, by the way, so those who think biofuels have a future past 2100 should think again.

Seventh, emissions take until 2100 to approach zero. There will be a long tail of both fossil-fuel-sourced kerosene and SAF biofuels into 2090 and 2100. Direct and indirect emissions are still substantial. Electricity is similarly going to have a long tail of lowering emissions. This is to be expected, and if not embraced, at least acknowledged. It's going to take a long time to get rid of the long-tail CO_2e emissions.

Eighth, hydrogen doesn't show up on this chart because the small amount of it that will be used to increase the yield of biofuel processes is assumed to be manufactured from green hydrogen with very low CO_2e per ton.

As mentioned in the discussion of hydrogen as an aviation fuel, the assessments of those who support it require an assumption of truly extraordinary advances that do not appear to be viable per the laws of physics. The economics of alternatives and operational requirements appear to make them so nonviable as to be easily dismissed.

To summarize, battery-electric aviation will be over 50 percent of flights by 2070 and approaching 100 percent by 2100. Biofuels will rise through 2060 and then start to fall again. This is an imperfect solution, but perfection is the enemy of good enough. This projection sees total aviation-related warming at under one-third of 2019 levels by 2060 and decreases that approach zero by 2100.

ADDITIONAL REFERENCES

https://www.baaqmd.gov/~/media/files/planning-and-research/research-and-model ing/saf-report-final-for-distribution-to-baaqmd-pdf.pdf?la=en?.

https://onlinelibrary.wiley.com/doi/10.1002/bbb.2225.

https://www.energy.gov/eere/bioenergy/articles/sustainable-aviation-fuels-low-car bon-ethanol-production.

https://www.flyradius.com/bombardier-q400/fuel-burn-consumption.

https://www.convertunits.com/from/gigajoule/to/gallon+[U.S.]+of+kerosene+type +jet+fuel.

https://theicct.org/sites/default/files/publications/CO2-commercial-aviation-oct2020 .pdf.

Chapter Twelve

Reinventing Buildings and Communities

BF Nagy

In the future, homes and buildings will be highly resistant to the fury of extreme weather and self-contained, collecting their own water and generating their own energy. They will be connected to a grid made up of other self-generating buildings, sharing energy to optimize community efficiency. Building envelopes will minimize energy loss to the outdoors.

Vehicles associated with building occupants will use energy generated by the building or by the vehicles themselves. Homes and vehicles will run more or less on 100 percent direct current (DC) power. Eventually, most industrial and institutional buildings and processes will do the same, except that "industrial" processes will have lighter footprints thanks to microelectronics, intelligent materials, and nanotechnology. Off-site energy generation and power utilities will fade. These homes and buildings will be built quickly on-site using 3D printing or will be 3D printed in factories as self-contained pods. Larger buildings will be assemblies of prefab pods and sections. Those who work in construction today, when they read this description, may think of reasons why it may never come to pass, but some of it is already happening. The construction business is notorious for being slow to adopt new ways; however, in large cities, significant change is already afoot.

In the modern world, the glaring advantages of new approaches are becoming too substantial to ignore. Digital design processes and virtual reality tools used by today's biggest developers, architects, mechanical and civil engineers, and contractors largely eliminate hours of duplication between design collaborators and provide precision and quality control as construction proceeds. Module prefabrication cuts construction schedules in half and increases quality. It has finally gained a solid foothold and is on a steady growth trajectory.[1] And 3D printing is poised to move beyond proof of concept to commercialization.[2] Due to economic benefits and the unstoppable penetration of electrification, these three technology groups and others will transform the industry.

Construction productivity will finally make significant gains along with building efficiency. Resilience and decarbonization will be designed in on day one. Solar panels will eventually yield to integrated solar cells; in other words, the windows and outside walls will collect energy without drawing attention to themselves and store it in capable batteries connected to electric heat pumps for space heating, cooling, and water heat for the bathroom and kitchen.

> "The construction business is notorious for being slow to adopt new ways; however, in large cities, significant change is already afoot."

We'll cook with induction, and the ventilation system will expel stale air and bring in fresh air while filtering out viruses and pathogens and effectively contain energy. These technologies mostly already exist and are proven, but adoption is ploddingly slow.

I've spent many years touring clean energy projects. In early 2019, I traveled down the West Coast visiting low-carbon buildings, micro-grids, and electric fleets in Washington State, Oregon, and California. In the near future, I'll be returning there and to Colorado and other Midwest destinations. In 2020 and 2022, I visited homes, buildings, shows, and facilities in twenty other, mostly eastern, states and five Canadian provinces. I've also reviewed projects in Mexico, Brazil, and several European countries.

During about twenty years as an environmental journalist, I've studied and witnessed at close hand the evolution of the technology landscape. There was a time when it was difficult to find people who were working on green buildings and transportation. It was unusual to find projects that went well technically or tracked data on economic viability. That's all changed. Green, clean-electrified buildings are plentiful and proven, although some of the tech described at the beginning of this section is still developing.[3] For example, 3D printing is technically feasible and economically viable, but its applications and case examples are limited and narrow.

Much of today's proven clean-building technology developed faster in Europe, where energy has always cost too much, or Asia, which has emerged as an economic power. But numerous important advances have more recently taken place in the Americas because, as expectations grew, the technology became more capable, more intelligent, and more digitally advanced. Relative efficiency improvements have been dramatic. This applies to every kind of structure at every price point: multi-family residential, affordable housing, rental apartments, commercial and retail properties, high-rise office buildings, and industrial facilities.[4]

In the past twenty-five years, global innovation and scalability in clean energy and clean-water technologies in the built environment have been revolutionary. And that's what we need—a revolution. Amazing developers, architects, and engineers are already reinventing our buildings for a new kind of world. They employ clean-energy and clean-water technologies such as rooftop and parking lot solar, home batteries, microgrids, air-source heat pumps, water-source heat pumps, ground-source heat pumps, heat pump water heaters, energy-recovery ventilators, wastewater or drain-water energy recovery, rainwater harvesting, smart-home or large-building control systems, and so on.

> "Developers, architects, and engineers are already reinventing our buildings for a new kind of world."

LOW EMISSIONS, HIGH RESILIENCE

When Hurricane Ian slammed into Florida in September of 2022, the Category 5 storm killed more than 140 people, injured thousands, blew away buildings, and knocked out power for about half a million households. It caused more than $112 billion[5] in damages, and many properties were inadequately insured or uninsurable.

Propelled by 150-mile-per-hour winds, a "two-stories-high" wall of water smashed through hundreds of homes and businesses in Fort Myers, where fourteen died. It destroyed other towns in the area, but one place called Babcock Ranch, about twenty-four miles from Fort Myers, was largely unaffected, even as the hurricane sat on top of it and blasted away at its defenses for nine hours. Even the community's hundreds of solar panels remained intact.[6] After his playing career, former National Football League offensive guard Syd Kitson created Babcock Ranch, a two-thousand-home, seventeen-acre development. He told me that he had seen raging storms off the New Jersey coast as a child and asked civil engineer Amy W. to create a town that was green, sustainable, and resilient in the face of extreme weather events.[7]

One Fort Myers survivor lost his best friend and told reporters that the water seemed to be about one foot deep at first, then in just a few minutes, it was nine feet deep. I visited Mayfield, Kentucky, which suffered massive damage in 2021. Witnesses there said it's unbelievable how powerful winds tossed cars, trees, and other heavy objects that crash into buildings and people.

Amy W. said, "Nature has survived for thousands of years. We figured out why it works and then copied it." Rather than bulldozing everything, they preserved the existing wetlands, ponds, and lakes for water storage and studied the existing flows. They raised the overall site by five feet and the hurricane shelter higher. "We used thicker pipe . . . and we did not lose water pressure or have main breaks. There were no surges in our water plants." Babcock Ranch had power, water, and internet throughout the storm. Residents could flush toilets and take showers, and few, if any, townspeople went to the hurricane shelter, only people from other communities. The town draws from a 150-megawatt community solar array and a 40-megawatt battery. Houses have sold quickly, and another eighteen thousand homes are planned. Amy W. said it doesn't have to be expensive to build better, especially if you plan up front. Babcock Ranch's selling prices are lower than the state average.

GEOTHERMAL AT WHISPER VALLEY

Some of the country's largest builders are partners at Babcock Ranch. This is true as well at Whisper Valley, one of several sustainable subdivisions in the Austin area in Texas. The latter will soon reach three thousand low-energy homes (eventually seventy-five thousand), all cooled and heated with geothermal. Whisper Valley emerged unscathed from the deadly Texas cold snap of 2021, which killed more than two hundred people with lower-quality homes and heating equipment not designed for freezing temperatures.

Geothermal is probably the best heating and cooling system so far invented in history, but adding the system to one house as a retrofit requires a relatively high up-front investment, because it consists of a geothermal field plus a ground-source heat pump for each home or building space.

Although initial costs are still high, in every project I've studied, ground- or water-source heating and cooling are among the least expensive options in terms of total life-cycle cost, which is how smart people with the means to plan ahead assess future systems. As mentioned, geo is more economical if it's integrated into the original design and if the geo field is shared among neighborhood homes, multiple units, or building sections and optimized with the help of today's intelligent load-management software. Chad B., the engineer for Whisper Valley, showed me that using smart software and linking the geothermal boreholes together as one community resource saves about 60 percent on energy.[8]

Ground-source heat pumps are more efficient than even the most modern air-source heat pumps and kick out the same high BTUs that you can derive from combustion appliances like gas furnaces, but heat pumps feed into the space slowly and quietly. When delivered via water-based radiators or other radiant devices, they create a luxurious, healthy, comfortable indoor atmosphere. Sometimes geo is combined with forced-air delivery, which still creates a better feeling than the gas-blast from our past.

Doug G., a construction veteran who conceived Whisper Valley, said, "The adjustments I've made as a traditional developer have not been dramatic. We're working with five major builders. . . . The changes we need to make in this industry can be embraced by all developers . . . all builders, and by all cities."

> "Ground-source heat pumps are more efficient than even the most modern air-source heat pumps and kick out the same high BTUs that you can derive from combustion appliances."

In Massachusetts, Connecticut, and New York State, Dandelion Energy has created an economically attractive model for retrofitting single-family homes, in cities, suburbs, and rural settings. The company has completed more than one thousand installations of its own ground-source heat pumps, with its own way to drill boreholes for geothermal and its own models to sell, lease, or finance for homeowners; with help from government incentives from the New York State Energy Research and Development Authority (NYSERDA).[9] One of its plans provides the installed system with no money down and a monthly payment of about $150 (in 2022).

At first, NYSERDA and the majority of New York State citizens were battling with gas companies over rates and pipelines, eventually wrangling these companies into partnerships in geothermal or other low-carbon community pilots. By 2022, they were proving successful, like others around the country.

Around the same time, large geothermal district energy projects were underway in Canada, such as the Blatchford community in Edmonton, Alberta, and Lakeshore Village in Mississauga, Ontario. The former will house thirty thousand residents downtown on re-purposed airport land using more than ten thousand heat pumps.[10] The latter may combine geothermal with heat recovery for twenty thousand residents. In Richmond, British Columbia, a geo district heating and cooling community was established in 2012, which keeps expanding. It's expected to eventually serve fifty million square feet.[11] Also in British Columbia, a wastewater heat-recovery residential/retail/industrial community has grown to thirty-seven buildings from a small Olympic Village built in 2010.

INNOVATIVE CONDO BUSINESS MODEL

For thirteen years, Tom W. struggled, with limited success, to sell geothermal cooling and heating systems to Ontario condominium developers. They kept asking him, why should they add to capital budgets with unfamiliar technology? Then it came to him: What if, like the gas companies, his firm became a utility?

He sourced a willing investment partner in Quebec, created a team to handle condo owner utility billing and maintenance, and, in 2015, went back to developers with a new message: His company would now own and maintain the geothermal systems in your condo buildings, reducing capital budgets and technology risks. It worked. There are now dozens of sizeable condo projects in Ontario and beyond, with geothermal fields under them.[12]

One of Tom W.'s first collaborators was Stan R., who manages a successful drilling firm called GeoSource, which works on condos and affordable multi-family, university, and government projects all over North America.[13] As a teenager, Stan planted hundreds of trees on his family farm, then later studied geology and engineering, and went to work in the Alberta oil sands. After learning about drilling deep wells, he returned to Ontario and started a geothermal company. This is what thousands of experienced oil and gas technicians and professionals are now doing, making the modern shift to low carbon, applying existing core skills to projects like geothermal.

HEAT PUMPS

Geothermal uses ground-source heat pumps, but there are other kinds of heat pumps, with the most affordable and ubiquitous being air-source heat pumps. Many people in North America think they are new because they have recently been popularized here, but heat pumps have been used for decades. Responding to market demand, heat pumps have advanced in the last ten years, performing better in cold and hot climates. In addition, as implied earlier, purpose-designed heat pumps and heat-pump-based hybrid systems have been introduced in recent years.

Electric heat pumps now outsell gas furnaces in the United States (in 2022 by 12 percent; 4,334,479 versus 3,872,368), after years of gains by the former and declining sales for the latter, according to the Air-Conditioning, Heating and Refrigeration Institute (AHRI).[14] In Europe, in 2021, heat pump sales surged by 34 percent, a reaction to high gas prices. After Russia invaded Ukraine in 2022, sanctions and gas-supply provocations made countries around the world more aware than ever before of the merits of energy independence. Heat pump sales in Europe increased by a further 38 percent in 2022 and have continued apace since then.[15]

> "Electric heat pumps now outsell gas furnaces in the United States."

Despite decades of global, widespread heat-pump success, fossil-fuel spin doctors pretend it is new technology that doesn't work and spread false news about how clean energy costs more or will overburden existing grid infrastructure. Evidence indicates otherwise.[16] In the 1970s, about 75 percent of homes in Stockholm, Sweden, were heated using oil. The government incentivized electric heat pumps. By 2014, about 8.5 million oil boilers were replaced. Greenhouse gas emissions dropped significantly. Although critics predicted disaster for Sweden's electrical grid, power use did not increase. It declined by about 22 percent due to efficiency gains.[17]

Fossil-fuel companies ignore this evidence and numerous studies that disprove their false narratives. Still, lazy or uninformed journalists in the United States and United Kingdom take the bait and talk about heat pumps as if they're experimental.

> "The government incentivized electric heat pumps . . . critics predicted disaster . . . power use did not increase. It declined by about 22 percent."

As mentioned earlier, electric ground-source heat pumps connected to geothermal fields (or water bodies) are as much as 400 percent as efficient as fossil-fuel combustion appliances. They provide the same high level of heat as the systems that burn things, polluting our air and costing us too much every month for energy bills. Modern air-source electric heat pump technology is about 300 percent as efficient as fossil appliances and at least twice as efficient as the old electric resistance devices like baseboard heaters.

TIGHT ENVELOPES

Because heat pumps feed heat and cooling into a space slowly, they work best in modern buildings with well-sealed envelopes, better windows, and more insulation. Efficient building envelopes are increasingly required in North American building codes. They are also part of the formula for net-zero and passive house models, which are increasingly common in building standards and codes.

Passive house (or *Passivhaus*) refers to a zero-carbon model incorporating technologies that achieve specific results, such as energy cuts as high as 80 or 90 percent compared to leaky twentieth-century buildings. For certification, passive house structures must perform as expected, based on measured outcomes (unlike early Leadership in Energy and Environmental Design [LEED] models). The key metrics relate to how tight the envelope is, temperatures, electricity, and emissions.[18] Because construction crews are accustomed to less precisely built homes, passive house is considered revolutionary in terms of quality.

SOLAR

Homeowners and business owners in the United States are adopting rooftop and other distributed solar arrays in huge numbers. As Dr. Audrey Lee writes in chapter 4, "It is expected that distributed energy storage and PV capacity across Oceania, North America, and Europe will double between 2020 and 2025."[19] In March of 2023, the Solar Energy Industries Association (SEIA) stated, "Nationwide, the residential segment installed just shy of 6 GW_{dc} in 2022, growing by a staggering 40% over 2021. A record 700,000 homeowners installed solar in 2022."[20] In the short term, solar (e.g., utility, microgrid, and single-building) is critical because it displaces dirty energy generated by gas and coal power plants.

AGRIHOODS ARE SPROUTING

Usually, I like traveling and visiting green projects, but in January of 2022, at Olivette Farm, it was brutally cold outside. The goats were huddling together for warmth, turning from a biting wind, whipping across snowy hills. It reminded me that even in North Carolina heating systems are needed in winter. I was glad to come inside and warm my numb fingers, while Scott A. explained the pesticide-free, farm-to-table food benefits and geothermal system requirements of a new development near Asheville.

Olivette's 4.5-acre organic farm provides monthly boxes of healthy food to seventy families who live there, plus some who don't. The 346-acre planned agrihood community is located on the French Broad River and includes trails, community gardens, a flower farm, community center, amphitheater, large private island, and fiber broadband internet. "From the beginning, we wanted to create a healthy place and honour the river. You're only going to build these houses once, so we ask homeowners to design for a HERS [Home Energy Ratings] rating of fifty-two with geothermal as the primary heating system. The architectural guidelines are in the deed . . . that's the only way to go."

By 2023, there were about one hundred developments in the United States that would be described as agrihood communities. They depart from the idea of homes heated with gas furnaces built around a golf course. The furnaces have become solar panels, batteries, and heat pumps, and the golf has become a family-friendly organic farm or orchard. During the COVID-19 pandemic, agrihoods provided community cocooning with a sense of safety.

AFFORDABLE LOW-CARBON HOUSING ·

The news on low-carbon affordable housing is also encouraging. Many of the green projects I visit are affordable or partly affordable multi-unit residential, and developments are growing larger. Connecticut is one of the smallest states in the United States, yet it is home to more than 140 beautiful state parks, including the Appalachian Trail, Campbell Falls, Mount Tom, and Gillette Castle. In these leafy sanctuaries, and indeed outside of them all over the state, citizens enjoy beaches, waterfalls, and trails filled with sycamore, black oak, maple, hickory, mountain laurels, asters, and red trilliums. Hikers catch glimpses of downy woodpeckers, green frogs, and snowshoe hares. In such an environment, young Chris B. found inspiration, donning her rubber boots on Earth Day and joining a community team in a Connecticut forest to help pick up litter.

> "Many of the green projects I visit are affordable or partly affordable multi-unit residential, and developments are growing larger."

She was shocked to discover that by summer's end people had thoughtlessly spread trash again in the same places: in the woods, in parks, and on the beaches. She was determined to be different, to help protect the beauty around her. Today, she's an architect specializing in low-carbon passive house buildings in New York State. Her firm is tackling one of the most difficult and critical challenges in our energy transition, retrofitting existing affordable housing, making it modern, green, and healthy, without evicting tenants and bulldozing sites.

There are many who say this work cannot even be done. It's easier to erect low-carbon single-family homes that are brand-new. Retrofits are difficult. Typically replaced at a rate of about 3 to 5 percent annually, our buildings may have a useful lives of a hundred years, meaning we're unlikely to replace them all in time for the worst of climate change. Most of New York City's larger, multi-unit fossil-fuel-heated edifices will still be standing in 2040. Affordable public housing tends to be pushed to the brink before modernization.

When we say the energy transition is going to be messy, late, and not perfect, we're often talking about building retrofits in cities. Constructing solar farms and producing batteries, electric cars, or solar panels, although complicated, pale by comparison to retrofitting millions of buildings. One engineer told me: "Every building is a custom job." It's not far from the truth. Most buildings still use double-whammy, dirty-heating systems powered by dirty electricity. Industrial buildings use a third emissions whammy: dirty-process operations. The built environment represents about 40 percent of the global emissions problem. Proportionally, it's even higher in cities.

In 2022, Chris B.'s team completed low-carbon conversions at Casa Pasiva, a nine-building, 146-unit project in Brooklyn's Bushwick neighborhood, while developing a model for healthy, cost-effective, deep-energy retrofits to occupied buildings that are certified to the passive house standard. It wasn't easy. The team had to fight for a zoning change that permitted eight inches of insulation on the outside of buildings. Tenants are not usually agreeable to moving out for a couple of weeks during renovations, but knowing they'll be able to control their own air-conditioning is motivating.

More insulation, along with better windows, heat pumps, induction cooktops, and energy-recovery systems reduces energy usage by 60 to 80 percent, significantly lowering operational expenses and emissions.[21] Chris B. is revered by the developer, Riseboro Community Partnership, which has more than one hundred more buildings to upgrade. "We're starting on them right after these," she told me. "Plus, a thirteen-story seniors' tower. We've been getting a lot of calls."

Suddenly in great demand, architects, engineers, and green-building professionals have survived decades-long struggles against industry acceptance, regulatory frameworks, undeveloped supply chains, and fear of change. In New York and cities all over America, gas and oil are quickly being phased out.

NEW LARGE GREEN BUILDINGS

If there is anyone left who doubts the scalability or commercial viability of heat pumps, net-zero, or passive house, they need look no further than Sendero Verde, an affordable-housing development of 709 units in Harlem, New York (or review bfnagy.com for fifteen-plus years of case examples).[22] Sendero Verde means "green path" and is among the signature achievements of two brilliant women: engineer Lois A. from Steven Winter Associates (SWA) and architect Deborah M. from Handel Architects.

The duo designed Sendero Verde within the confines of an affordable-housing budget, and their companies also collaborated on The House, a twenty-six-story passive house–certified student residence for Cornell University's tech school on New York's Roosevelt Island. Another of their projects is the massive fifty-three-story Winthrop Center in Boston, which, in the early 2020s, was considered the biggest passive house office project in the world.

The two-building Sendero Verde complex has double-pane windows, an envelope at least five times tighter and better insulated than typical buildings, and is heated and cooled by a commercial-scale heat-pump system. Engineers combined an innovative solar-gain approach to heating/cooling zones with intelligent modern control systems to optimize energy and piping.

"All our projects are at least 50 percent lower emissions. . . . This is the way that projects will be built."

"We have the data now. All our projects are at least 50 percent lower emissions," said Dylan M., a young passive house consultant at SWA.[23] He says climate change is the single most important issue for the United States right now. "Things are going in the right direction, but I'm not sure it's fast enough. I worry because I want to have kids one day."

FIFTY-THREE-STORY PASSIVE HOUSE OFFICE BUILDING

Millennium Partners is one of the biggest construction companies in the world. In January of 2022, when we were working on a film on the thirty-second floor of the aforementioned Winthrop Center in Boston, the company's VP, Brad M., commented on zero-energy buildings: "This is not just a blip. This is the way that projects will be built. . . . We're never going back to the old way."

The 1.8-million-square-foot Winthrop Center is a huge office, residence, and shopping mall that will emit about 150 percent less carbon than a typical Class A modern high-rise and 60 percent less than LEED Platinum.[24]

PREFAB PASSIVE HOUSE

Lois A., Deborah M., and Dylan M. also worked on the aforementioned student residence at Cornell University, which became a learning lab for prefabricated passive house panels. These are super-insulated wall sections, sometimes with windows and doors already built in, that are manufactured in advance in a factory. Prefab helps ensure adherence to the passive house standard and makes it faster and easier to build when necessary. The Cornell project was located in a tight spot with other existing buildings around it.

In some circumstances, prefab is not more expensive and offers other advantages. Prefab companies have finally begun to enjoy rapid growth, finding the right niches, such as hotel room pods and hospital bathrooms. Design digitization pairs well with prefab and can save months or years on tight construction schedules, avoid bad weather, and improve safety for workers. Modern off-grid living is also returning, in the form of small self-contained pod-like structures with integrated solar and water systems, extremely efficient envelopes, and high-tech features.[25]

Prefabrication can become significant in unexpected ways. In the 2020s, the UK government announced a heat-pump/insulation program that was beset by budget overruns and quickly cancelled. This was followed by musings within the industry on more cost-efficient approaches to retrofits that might quickly be scaled up for thousands or millions of existing drafty homes with oil boilers.

"Bolt-on" solutions were floating around on the internet, designed for small village houses, row housing, and so on. One incorporated a heat pump and ventilator built into a vestibule that could be added in front of a main entrance, while improvements to windows and envelope leaks reduced energy loads. Governments are now desperate for this kind of one-size-fits-all solution.

ENERGIESPRONG AND SIMILAR MODELS

Energiesprong is a retrofit standard that originated in the Netherlands and expanded to France, the United Kingdom, Germany, and the state of New York.[26] A new insulated cladding is installed to the exterior of a building, along with a new insulated roof over the existing roof. The new cladding and roof are air-sealed together and gasketed windows and doors replace the old ones. It only takes one week to complete and comes with a thirty- to forty-year warranty. In New York, existing gas systems are replaced with air-to-water heat pumps, which supply hot water for space heating, bathing, and cooking. Existing radiators are sometimes retained as part of the new heat-pump system.

In the early 2020s, Amory Lovins's Rocky Mountain Institute asked progressive Philadelphia architect Tim M. to adapt Energiesprong for its deep-energy program known as REALIZE. Tim's group began assembling US design teams, developing clients, and researching local prefab suppliers. Within a few years, they were already working on deep-energy retrofits in the Massachusetts area and a large project in Louisville, Kentucky.[27]

AFFORDABLE LOW CARBON IN PHILADELPHIA

Years before REALIZE and the Inflation Reduction Act, Tim M. and Ron C., his solar engineer, were using the combination of air-source heat pumps, tight envelopes, and solar to great effect on numerous affordable housing projects in Philadelphia. Then they went further, walking officials at the Pennsylvania

Housing Finance Agency (PHFA) through a passive house (PH) project they had completed for about the same cost as other developers were budgeting to meet the less-ambitious building code. The PHFA was intrigued. As a pilot test in 2015, it offered potential affordable-housing-funding recipients ten optional points for PH designs. This attracted attention, and 38 percent of applications committed to PH.

The PHFA funded eight PH projects, creating 422 new affordable-housing units at $169.00 per square foot, about 2.5 percent higher than $164.50 for code-built projects. In the second year, the PHFA funded ten PH projects, but the numbers flipped. PH projects came in for 1.8 percent LESS cost, at $167.50 per square foot versus $170.50.[28] The builders had learned how to build better for less. The trend has continued since, with PH buildings costing about the same amount as leakier, higher-carbon buildings.

ARE THEY BETTER?

According to PHFA, utility costs are much lower. A veterans' housing project of forty-nine apartments completed in 2017 estimated total annual utility costs at $27,372, compared with housing agency allowances of $64,548, thus saving about 58 percent. The Department of Energy says PH lowers energy costs around the country by 50 percent.[29]

Because of efficient heat pumps, on-site emissions are significantly lower; however, much of Pennsylvania's electricity is still generated using gas and coal. As part of its goal to cut carbon emissions by 80 percent, Philadelphia is working on 100 percent renewables for municipal operations, and in 2024, a new solar plant will supply 22 percent of the government's power.

Pennsylvania's experience is being repeated around the United States. My files include numerous PH or net-zero affordable-housing projects that cost about the same to build, or a few percentage points more, and much less to operate. In 2018, the Massachusetts Clean Energy Center (CEC) funded eight PH affordable-housing projects, five of which were complete and operating by mid-2022. CEC reported up-front incremental costs under 3 percent, and they inspired a groundbreaking $4 billion state energy efficiency plan in 2022 created by the Massachusetts Department of Public Utilities.

> "My files include numerous passive house or net-zero affordable-housing projects that cost about the same to build."

The Passive House Institute US (PHIUS), working with the aforementioned Pennsylvania group's scheme, has approached forty state housing finance agencies and encouraged them to offer points for the PH model. By 2023, eighteen had committed to do so, and fourteen more were seriously considering it. Many on the list are red states, so clean energy may be returning to bipartisan-issue status in the building sector, too.

HEAT-PUMP WATER HEATERS

The required water temperatures are higher for bathing and cooking than for space heating. For a time, this was a somewhat-viable rationale for continuing to use fossil-fuel combustion heat for domestic hot water, but technology improved and extreme weather became more frequent.

In the early 2020s, modern, highly capable heat-pump water heaters were introduced and rediscovered by North American building professionals, who began avoiding gas by installing them in large numbers, even as they were still a little too costly to create a compelling payback narrative. With volume driven by decarbonization mandates, growing product efficiencies, and occasional spikes in gas prices, the payback numbers are improving quickly, and municipalities are phasing in bans on fossil fuel for domestic water heat, too. In addition, people like those on Tim M.'s team are employing innovative configurations.[30]

ELECTRIC INDUCTION COOKTOPS

One of the mythologies that the gas industry has worked hard to perpetrate is that electric-induction cooking is either difficult, expensive, or not fulfilling for those who prepare food. As usual, over time, the truth is coming out. In 2023, *Chatelaine* magazine interviewed chefs who love induction stoves.[31] One big-city restauranteur said induction cooking was simple because you set it and it remains precisely at the specified temperature. The expected big electricity bills never materialized, perhaps because it was easy for his cooks to be more efficient. Another had moved to a small town where she disliked the existing propane stoves that burned too hot for her carefully simmered, delicious soups. Another was a baker and said the syrup for her butter cream can be made faster with induction. She also likes precision temperature control and auto-shutdown.

Cooking with gas or wood creates emissions and pollutes indoor air, emitting nitrogen dioxide, carbon monoxide, and twice as much particulate matter as electric or induction stoves.[32] Increasing medical evidence links cooking with gas and respiratory and cardiovascular health difficulties.[33] Some 12.7 percent of childhood asthma can be attributed to gas stove use,

according to a recent study published in the
International Journal of Environmental Re-
search and Public Health.[34]

With induction, the cookware and food are
heated, but the remainder of the appliance
remains cool, so there is less risk of burns
and fires. Spills don't get baked on, making

> **"Increasing medical
> evidence links cooking
> with gas and respiratory
> and cardiovascular
> health difficulties."**

cleanup easier. Modern induction stoves are convenient and attractive, more
efficient, and the initial investment is not substantially higher.

According to the Environmental Protection Agency (EPA) in the United
States, estimated greenhouse gas emissions from non-renewable fuels for
cooking amount to one billion tons of carbon dioxide per year.[35] The World
Health Organization (WHO) says that around the world, some 2.4 billion
people cook using open fires or inefficient stoves fueled by kerosene, biomass, and coal, generating unhealthy household air pollution.[36]

HEAT PUMP CLOTHING DRYERS

Heat pump clothing dryers (HPCD) increasingly make sense; however, they
are not usually at the top of the list of climate actions. It's more important to
ensure your building's heating, hot water, and power functions are decarbonized than investing in better clothing dryers. A large part of the world doesn't
use clothing dryers. They hang their laundry in the breeze. Hanging and
HPCDs take longer, but reduce wear and tear when compared with high-heat
dryers. As of 2023, HCPDs still cost more up front but save at least 50 percent on energy. Incentives to help with the original purchase may be available
from a local government or power utility.

SMART HOMES AND BUILDINGS

A peculiarity in the clean-energy/clean-water technology transition is the
rarely discussed but highly significant impact of equipment and process digitization and electronics. Like the computers they work on, IT professionals
can seem to be from another world, but when they land on the surface of the
Earth, we often realize that they are amazing. As mentioned, at Whisper Valley, the optimization of a geothermal network using intelligent software saved
about 60 percent on energy. No small thing.

At a school board meeting in Ontario, Canada (a country that seems to effortlessly spawn a continuous flow of brilliant IT people), the facilities team

decided twenty years ago to completely digitize the operation of 105 of its schools. They applied for government grants that allowed them to hire coders under numerous projects and spent two decades on the most impressive application I've ever seen of a building automation system (BAS).

They have achieved cost savings of 47 percent and an improvement in weather-adjusted energy intensity to 50.12 megajoules/square feet in 2015 compared with 89.4 megajoules/square feet during 2000. In dollar terms, this is $4.5 million in unnecessary cost avoidance. The director of facilities can pull up a dashboard that shows the real-time operating status and energy consumption of every device in every school in their network.

To connect just one school involves about three hundred data points. A point is one function on one piece of equipment. The temperature of water in a heat pump is one point. Changing the speed of a fan is one point. Making sure it changed is another point. Each of these points requires mechanical devices, sensors, and so on physically in the school, plus programming into the system. To add one school to the system could take four people two to three months.

HERITAGE OFFICE BUILDING

The Sun Life building in downtown Montreal has twenty-six floors and totals more than one million square feet. Over the years, the management team has replaced its older systems with more than sixteen hundred water-source heat pumps and energy-recovery ventilators, which manage 95 percent of cooling and heating. It is a masterpiece of zone management, using sophisticated computer controls to recover and move heat and cooling throughout the structure so efficiently that surprisingly little active HVAC is required. "The interior always needs some cooling, and the perimeter usually needs some heating," says Operations Manager Pierre P. "We have a lot of heat-recovery opportunities. Our tenants are high-tech companies with data centers and numerous employees. We draw heat from those zones."

The science and engineering of smart systems controlling energy recovery, energy movement, and energy intensification is beautifully demonstrated at the Biodome, an indoor zoo, also in Montreal. It is home to forty-five hundred animals and five hundred kinds of plants, living in four different climates—a cold polar environment, a hot tropical rainforest, and two additional moderate-temperature exhibit areas. Engineers recently replaced an old fossil-fuel steam system and ancient refrigeration gear with a modern water-source system that draws energy from a naturally occurring aquifer located sixty feet below the building and saves $980,000 each year. The 52 percent energy saving won the Biodome an Association of Energy Engineers (AEE) award and an American Society of Heating, Refrigerating and Air-Conditioning Engineers (ASHRAE) award.

When the outside temperature is between 23°F and 32°F, the water-source system is not needed. Instead, sophisticated controls automatically shift energy among the four areas and pool temperatures are adjusted to match

"Sophisticated controls automatically shift energy among the four areas."

air temperature requirements. The air is cleaner, and reduced volumes are needed through the use of energy recovery ventilators (ERVs). More than 80 percent of heat energy is recovered and recirculated between zones. Equipment control sequences are optimized for peak demand, time-shifting, sensor data, and calculations of the effects of pools as energy sinks. Efficiency has been increased using the ERVs and variable-speed pumps and fans.

ENERGY RECOVERY

There are numerous ways to reduce, reuse, or recover energy that our equipment and older building envelopes manage to waste. Most of these are covered in the refrigeration section of this book, some in sections on data center energy and wastewater energy. These technologies are all well-proven and surprisingly effective for deep cuts in energy use. The same can be said for heat recovery ventilators (HRVs) and energy recovery ventilators (ERVs), which pair perfectly with heat pumps in many climate zones and offer 100 percent fresh outdoor air with minimal energy loss and pathogen filtration possibilities. The majority of PH projects I've seen include an HRV or ERV.

GREENING DATA CENTERS

In the late 1990s, the IT industry realized that energy use for computing was on a steep upward trajectory and that the industry was on a collision course with environmentalists. Data centers (DCs) consume vast quantities of power for both processing and DC cooling. Amazon, Google, and Microsoft, the big three cloud providers, went to work testing numerous ways to reduce power use for these two key functions and to source clean power for their operations.

They began locating data centers in northern regions where outdoor air was cool enough to be brought in for "free-cooling." They ran virtual machines on their servers to limit downtime, installed custom cooling systems, automating wherever possible. Microsoft even experimented with tossing a container full of servers into the ocean to see if that would reduce cooling energy. After numerous pilot tests, data centers have begun to adopt liquid-cooling systems because liquid cools more efficiently than air. It sounds tricky given that liquids and electricity don't generally mix; however, innovative solutions have been created.

These initiatives achieved good results with power usage effectiveness (PUI) ratios, and despite significant increases in global computing between 2005 and 2015, its total electricity consumption remained fairly constant. There are thousands of data centers in companies around the world that may or may not be purchasing cloud-computing services from the big three and/or undertaking these kinds of initiatives for their own server facilities; however, when they did move their data from in-house servers to the cloud, they often reduced their energy consumption and emissions footprints.

At the same time, the world's electricity was becoming greener, and in recent years, the industry has concluded that social media and artificial intelligence (AI) activity is likely to continue to increase, so clean power must become the focus. The big three have gone beyond locating key DCs in close proximity to renewable power sources. They are now building renewable generation on a mass scale. In 2021, Kara Hurst, Vice President of Worldwide Sustainability at Amazon, said, "We reached 65 percent renewable energy across our business in 2020 and became the world's largest corporate purchaser of renewable energy . . . we are on an accelerated path to power our global operations with 100 percent renewable energy by 2025."

Amazon is huge, with eighty-one availability zones in which its servers are located and operations in 245 countries and territories; however, it is still likely that it will meet the goal because it's now a key performance measure that affects earnings. Clean energy is cheaper than dirty energy and brings with it a host of efficiency-improvement opportunities. Amazon has already allocated the funds and is busy acquiring and building renewables projects. Corporations and data center managers around the world who are still managing their own servers should look to the leadership of Amazon, Google, and Microsoft and go green.

THE NEXT BIG THING IS SEWAGE

Believe it or not, the next big thing in clean energy will be the untapped wealth in sewage. Denver's National Western Stock Show and Rodeo has secured rights and is updating about one million square feet of new indoor spaces, 90 percent of which will be heated and cooled with "free" energy from mostly existing local sewer trunk lines. Denver is also promoting its wastewater energy resources with additional customers around town. Other North American cities are studying the potential. The technology now exists to make a compelling economic case for wastewater energy recovery in almost any municipality. For those seeking high-impact ways to meet greenhouse gas reduction targets, this is simple, proven technology.

I recently toured DC Water's new head office in Washington. Saul K. showed me around and explained how easy it was to heat and cool 170,000 square feet of office space using recovered wastewater energy. The principle is the same as geothermal, except the resource is city sewage water rather than below-ground earth. In both cases, the resource is cooler than above-ground air in summer and warmer in winter. You intensify the difference using purpose-built heat pumps.

> "The technology now exists to make a compelling economic case for wastewater energy recovery."

DC Water's beautiful modern headquarters has been constructed atop an existing sewage treatment plant. A heat exchanger extracts thermal energy from wastewater and uses it to heat or cool a separate stream of clean fluid. Heat pumps intensify and distribute the heated or cooled fluid through the building in the same way as a conventional boiler and radiator system would. It's efficient, and the energy resource is already built. You simply tap into it.

Lynn M., the CEO of a Vancouver equipment manufacturer, says DC Water saves about 35 percent on air-conditioning and 85 percent on heating, plus four to six million gallons of water each year that would have been used by the cooling tower if they were not using wastewater energy instead. Saul K. said the system was sized for the building on the hottest and coldest days, reaching about forty megawatts monthly in July and August, but the wastewater resource available at the site is far bigger than what is being used.

"It might be able to do ten buildings of this size, or more." DC Water has been talking to potential customers about this kind of expansion. The project helped DC Water achieve operational cost savings and greenhouse gas reductions. The American Geophysical Union installed a similar system at its headquarters, also in Washington, DC.

European and American studies on the potential of wastewater energy reveal significant potential. In Switzerland, it could provide 7 percent of the country's heating demand. In Germany, it could heat and cool two million homes. A New York City study concluded that if 5°F of heat were removed from wastewater flowing through the sewer pipes beneath the streets over the course of one year, $90 million worth of energy could be recovered.

VANCOUVER, BRITISH COLUMBIA

The biggest operating wastewater project in North America is very likely the False Creek district heating facility in Vancouver, Canada. During an urban planning webinar in 2021, my guest manager Linda P. recounted that phase one was built to heat the Olympic Village in Vancouver and two other buildings, totaling about 1.5 million square feet. In 2021, this increased to 6.5 million square feet. Two large heat pumps supplied seven residential buildings, all of Science World, some campus buildings at Emily Carr University, a large Mountain Equipment Co-op store, a community center, and an office building. "At full build out, we will be serving thirty-seven buildings, totaling 21.5 million square feet," she said. North American families' showers, cooking, cleaning, laundry, and toilets send an average of about three hundred gallons of water down the drain each day.[37] It represents an opportunity to draw untapped energy from more than thirteen trillion gallons per year.

Chapter Thirteen

Hockey, Natural Refrigerants, and Greening Industry

BF Nagy

In 2023, the Vegas Golden Knights, the self-described "misfits" of the National Hockey League (NHL), won the Stanley Cup, proving that a new team made up of players left unprotected by other NHL teams could beat everyone in the world's top hockey league. In the same year, an even newer team, the Seattle Kraken, made it to the playoffs and defeated the defending Stanley Cup–winning Colorado Avalanche. For Seattle, it was just their second season in the league. They're young, fast, fresh and environment friendly. In fact, their home ice is called Climate Pledge Arena because it incorporates numerous green technologies and practices.[1]

Today's arena management needs to be climate aware because their facilities use loads of costly energy and resources, and major league sports is facing revenue challenges. One seldom-told story is that quite a few NHL arenas and practice facilities are now moving quickly in directions that are better for the environment. An engineer who has worked on more than two hundred arena projects told me that systems using heat-recovery technologies and natural refrigerants have been installed by the St. Louis Blues, San Jose Sharks, Anaheim Ducks, Arizona Coyotes, and Ryerson Rams at Toronto's Maple Leaf Gardens/Mattamy Centre. More NHL teams are expected to follow.[2]

WHAT IS ENERGY RECOVERY?

If you pulled an old model refrigerator out from the wall and stood behind it while it was running, you would probably feel heat blowing on you. That's the heat being removed from inside. You can say refrigerators create cold, or you can say they remove heat. It's the same principle used by the humble heat pump, now touted as a world-changer in space heating and cooling. Most of the heating or cooling from an electric heat pump is not so much generated by the device as it is moved from another source and intensified.

In arenas, modern heat-recovery equipment captures the heat that would otherwise come out of the ice-making equipment and be expelled outside the building and instead channels said heat to the snack bar or other common areas to improve our comfort. Modern grocery stores now do this, too.

WHAT ARE NATURAL REFRIGERANTS?

They are proven replacements for the synthetic gases that helped with the physics of moving and intensifying heat energy and cooling energy but became especially environment-damaging after they were introduced about fifty years ago. Refrigerants are supposed to be sealed inside of refrigeration equipment most of the time, but studies show they leak out a great deal[3] and can be very dangerous. Addressing harmful refrigerants was ranked among the highest priorities in an important climate book titled *Drawdown*, which was edited by Paul Hawken and published in 2017.

Refrigerants are environmentally perilous in two ways. The original Montreal Protocol was created due to a hole that opened up in the ozone layer of Earth's atmosphere above the South Pole. This atmospheric layer helps keep the sun from burning us all to a crisp. Scientists indicated the hole would continue to expand if the use of damaging refrigerants was not curtailed. The world came together, and 197 countries signed the Montreal Protocol in 1987.[4] This resulted in about thirty years of replacing hydrochlorofluorocarbons (HCFCs) with hydrofluorocarbons (HFCs). And it worked. The ozone hole stabilized and then began to shrink.

GLOBAL WARMING

The Kigali Amendment to the protocol recognized that although HFCs were better for our ozone layer, they contributed significantly to climate change. So again, the world's countries came together and agreed on an amendment that this time they would phase out HFCs and phase in natural refrigerants, used in modern equipment that was better designed for them.

Chemical companies met with the Trump administration, which then attempted to block ratification by the United States of the natural refrigerants amendment. But there was intense pressure, both international and domestic, and in late 2020, bipartisan legislation known as the AIM Act was introduced to phase down HFCs. The Senate later ratified the Kigali Amendment with strong bipartisan support, joining 150 other countries.

This wasn't motivated only by concern for our home planet. America will benefit by re-asserting its economic leadership. Equipment manufacturers were worried they would lose market share and venture capital to foreign companies in the absence of harmonized standards. New kinds of systems were penetrating the market from smart innovators in Quebec, Europe, and Japan (and the United States). Distributors wanted a simplified supply land-scape. Technical trade groups and installers wanted national rules, not state patchworks, and finalized training directions.

When federal governments politicize or delay global policy efforts, senior civil servants become anxious. In this case, several years of foot-dragging led to some state governments passing their own laws, potentially leading to a messy patchwork situation that can take years to clean up after the political gamesmanship ends.

The AIM Act directs the EPA to phase out HFCs by 85 percent by 2036. Chemical companies opposed it because new technology works with natural refrigerants such as CO_2, ammonia, and propane rather than expensive blended refrigerants the chemical companies developed. Synthetics mostly still have high global warming potential (GWP) or ozone depletion potential (ODP), or both.

In the world of refrigerants, ammonia, CO_2, and propane, with relatively benign GWPs of zero, one, and three, are good guys. They're being used in relatively small amounts inside refrigeration equipment, but with leakage, GWP matters.[5] The EPA estimates that cumulative net benefits will be $272.7 billion from 2022 through 2050.[6]

NATURAL REFRIGERANTS

Examples of successful implementation with natural refrigerants continue to multiply; they are found all around the United States and in other countries. In the last few years, I've published articles and hosted webinars on projects such as the aforementioned NHL hockey arenas, small town arenas, big grocery store and warehouse systems, and hundreds of thousands of smaller R290 propane-based standalone refrigeration cases now being used world-wide on the floor in retail stores by Coke, Pepsi, Unilever, Nestle, McDonald's, Aldi, Whole Foods, Target, and Starbucks.

HillPhoenix of Georgia has supplied bigger CO_2 transcritical systems to hundreds of grocery stores across North America, including numerous Walmarts, and in warehouses and food plants. The AHR Expo, the world's biggest annual heating, cooling, and refrigeration event, has given Danfoss and other firms some of its top innovation awards for CO_2 systems and other green products in the last few years.[7]

REFRIGERANT	ODP Ozone depletion potential	GWP Global warming potential
NATURAL REFRIGERANTS		
R-717 Ammonia - NH_3	0.0	0.0
R-744 Carbon Dioxide - CO_2	0.0	1.0
R-290 Propane	0.0	3.0
R-718 Water - H_2O	0.0	0.0
R-729 Air	0.0	0.0
SYNTHETIC REFRIGERANTS		
R-123 Dichlorotrifluoroethane	0.02	0.02
R-114 Dichlorotetrafluoroethane	1.0	3.9
R-502 (48.8% R-22, 51.2% R-115)	0.283	4.1
R-11 Trichlorofluoromethane	1.0	4000.0
R-12 Dichlorodifluoromethane	1.0	2400.0
R-13 B1 Bromotrifluoromethane	10.0	6290.0
R-22 Chlorodifluoromethane	0.05	1700.0
R-32 Difluoromethane	0.0	650.0
R-113 Trichlorotrifluoroethane	0.8	4800.0
R-124 Chlorotetrafluoroethane	0.02	620.0
R-125 Pentafluoroethane	0.0	3400.0
R-134a Tetrafluoroethane	0.0	1300.0
R-143a Trifluoroethane	0.0	4300.0
R-152a Difluoroethane	0.0	120.0
R-245af Pentafluoropropane	0.0	1030.0
R-401A (53% R-22, 34% R-124, 13% R-152a)	0.037	1100.0
R-401B (61% R-22, 28% R-124, 11% R-152a)	0.04	1200.0
R-402A (38% R-22, 60% R-125, 2% R-290)	0.02	2600.0
R-404A (44% R-125, 52% R-143a, R-134a)	0.0	3300.0
R-407A (20% R-32, 40% R-125, 40% R-134a)	0.0	2000.0
R-407C (23% R-32, 25% R-125, 52% R-134a)	0.0	1600.0
R-507 (45% R-125, 55% R-143)	0.0	3300.0
R-458A Bluon TdX 20	0.0	1564.0
R-1234yf (Vehicle A/C)	0.0	1430.0

Figure 13.1. Natural Refrigerants vs Synthetic Refrigerants
Figure created by Climate Solution Group, using data from Engineeringtoolbox.com.

Shecco, an international refrigeration consultant, says that transcritical systems in grocery stores allow better temperature control, cost 10 percent less to install, use 15 percent less energy, reduce building heating costs by 75 percent through heat recovery, and reduce maintenance costs by up to 50 percent.[8] They're just newer, better systems.

A couple of years ago, the president of a global refrigeration firm told me: "CO_2 is nontoxic, not flammable, naturally occurring, causes no ozone depletion and costs about a dollar a pound. Synthetics are $40 per pound. There are now about 30,000 transcritical CO_2 systems around the world."

NEW TECHNOLOGY DEVELOPMENTS

Another system has emerged from early pilots that solves some problems: CO_2 combined with low-charge ammonia. The system is configured to keep the ammonia in the machine room and use the CO_2 in areas closer to people, optimizing safety and system efficiency. By 2020, there were already more than two thousand installations in Europe, more than nine hundred in the United States, a couple hundred in China, and another nine-hundred-plus in Japan.[9]

Writing on Manfacturing.net,[10] Chuck Taylor and Todd Allsup gave the example of a typical two-hundred-thousand-square-foot refrigerated warehouse using a conventional central station ammonia system with a refrigerant charge of around forty thousand pounds of ammonia. The same facility combining CO_2 and ammonia would require a refrigerant charge of less than seven thousand pounds. This latter level is below the threshold that triggers requirements for compliance to process safety management (PSM) and with the Department of Homeland Security.

According to Freor.com, R290 propane systems offer 30 percent operating savings compared with now-obsolete HFC systems. For rack refrigeration, CO_2/ammonia systems have a potentially higher up-front cost than HFC systems, but they also have lower operating costs, resulting in a three- to five-year payback on the investment.[11]

Mayekawa, a Japanese industrial refrigeration firm, has been creating cutting-edge, environment-friendly solutions for decades. In one case, there was no gas service in the region of a northeast seafood plant, and operators were suffering the high cost of oil-fired water heaters. "They already had a regular ammonia system for making solid ice blocks. We took the gas from that system before it was rejected outside and increased it to 450 pounds pressure. Now they have plenty of far less expensive hot water at 158°F without fossil fuels," said Colin J. of Mayekawa.

Mayekawa also invented PascalAir, a system based on the principle that air generates heat when it is compressed and loses heat when it is expanded; however, while expanding, its temperature drops below the original temperature. They've installed more than one hundred PascalAir units for low-temperature tuna storage and in the pharmaceutical industries.[12]

Natural refrigerants for heat pumps, air-conditioning, deep-freeze food plants, grocery stores, and retail food cases will soon become the standard for refrigeration, space cooling, and heating. In addition, CO_2 heat pump water heaters for residential and commercial purposes are now being installed in North America.

This progress in the cooling and refrigeration sector reflects a hard-won global consensus. It proves we can agree on best solutions, ignore short-term, self-interest, and fabricated politics, take action, and achieve results.

But it has required thirty or forty years, and there is still much to be done. We no longer have such generous timelines. The planet is on fire. Let's draw inspiration from the refrigeration industry, which is moving full speed ahead into a new era of lower-cost, environment-friendly cooling and refrigeration systems—and ice-making for hockey.

GREENING INDUSTRY

Modern equipment using natural refrigerants is well-proven, but technology for green cement and green steel, other industrial processes, long-haul shipping, and rail and air travel are true laggards in the world of decarbonization. They're finally advancing a little more quickly. For some of these sectors, status quo industries have promoted hydrogen as a solution, but unfortunately, more hype than substantial progress has materialized over several decades. Conventional concrete production and use creates 8 or 9 percent of all greenhouse gas emissions. Steel is also a significant emitter.

Green Concrete

A company in Rives de l'Yon on the west coast of France appears to be well on the way to commercializing low-carbon concrete using three key inputs: blast furnace slag from the metallurgical and steel industry; flash clay, a co-product of clay sludge; and gypsum/desulfogypsum from construction site excavation. Apparently, the three are abundant and easily sourced. They are used to create cement products that seem to work for most of the traditional applications. The company says they're green, and there is no clinker production involved, which apparently creates most of the emissions. Nothing is burned, and no quarrying of raw materials is needed. The firm seems to have financial backing, a partnership with Euro industry leader CEMEX, sound manufacturing plans, and is developing distribution arrangements. There are other companies on both sides of the Atlantic working on green cement solutions, a hopeful sign for the future.

Green Steel

At least one firm in Sweden has produced some steel that it calls green and which has been used by Volvo, the car and truck manufacturer, in a large prototype vehicle that carries ore during mining operations. The process involves hydrogen as a replacement for fossil fuels in the iron pellet and iron manufacturing process and the use of an electric arc furnace to transform the iron into steel. The company says the energy to create the hydrogen, via fuel cells, will eventually come from renewables, presumably wind and solar. The skeptical language here is based on similar claims made in other heavy industries, which have not yet come to fruition. With hydrogen projects, the available technical information is often limited and murky and the timelines to commercialization mysteriously long, but political "research" partnerships are magically in place early, with public money flowing.

MOTORS

There are millions of motors used in the United States and around the world. A company in Round Rock, Texas, says that its efficient modern motors can save about half on electricity. The patented design eliminates much of the iron and copper and operates via etchings on circuit boards. "It's eighteen pounds versus eight pounds. . . . Smaller, quieter, lighter, same output, higher current density." The company has succeeded commercially, is winning product awards, and is now developing the product line, creating motors in numerous sizes for different applications.

THERMAL BATTERY

Some industry greening has been described as impossible by industry personnel, for example very high temperature systems. A group in Fremont, California, is attempting to ramp up production of a heat battery that offers a zero-emissions source of industrial heat, storing solar and wind energy at temperatures above 1,200°C. This might be a better route than the aforementioned hydrogen journey. The new heat battery is designed to pull energy from solar, wind, and the grid, charging brick materials intermittently, while delivering continuous heat. The company says it will enable industries such as cement, fuels, food, and water desalination to reduce emissions and leverage the falling costs of renewables.

Chapter Fourteen

Polar Biodiversity and Climate

David K. A. Barnes, PhD

David Barnes, PhD, of the British Antarctic Survey, is a marine ecologist who spends months and even years on research stations and vessels in the Arctic and West Antarctica. In fact, he sent his article for this book from Antarctica. He is an advisor to the Intergovernmental Panels on Climate Change (IPCC) and Biodiversity and Ecosystem Services (IPBES), has been part of more than twenty polar research expeditions, serving as leader on at least six of them, and was lead author on the majority of some 298 peer-reviewed publications, which have been cited more than twenty-four thousand times. Winner of a 2023 Polar Medal and the 2023 APECS international mentorship award, he has also contributed as author/editor to several books and given more than seventy television, radio, and news interviews in fifteen countries.

The very serious climate crisis we face is intertwined with one of drastic nature loss. These crises influence each other, but the way we try to mitigate them so far rarely reflects that. This has been to the detriment of both. Yes, rapid, urgent, and massive change is needed, but it is crucial that this is synergistic across these problems. Action to limit, and ultimately reverse, climate change requires drastic strategies in each of major contributory sectors (e.g., construction, travel, etc.). However, climate amelioration strategies which also consider impact on nature are generally more effective and vice versa. Nature-based solutions may be a small part of the solutions palette, but it is a very important one because of its efficiency, cost-effectiveness, and ease to implement—in other words, it is a "low-hanging fruit."

For example, a variety of different habitats can capture and bury carbon more efficiently than human technology.[1] This process is greatest in near-intact (undamaged) wet habitats, where protection goals are long-term and holistic, factoring in everything from ecosystem to human health. That

129

Figure 14.1. Barnes's ship.
Courtesy of David Barnes

might sound like a daunting list, but it need not be; it just means identifying and prioritizing carbon- and species-rich hotspots and protecting these in a sensible manner. We need to make sure such protection is scientifically underpinned to actually mitigate against identified threats, rather than parks "only on paper." Well-meaning and attention-grabbing strategies like 30 by 30 (protection of 30 percent of land and sea by 2030) would be much more effective if areas designated are targeted to biodiversity functionality and threat mitigation.

> "Nature-based solutions may be a small part of the solutions palette, but it is a very important one because of its efficiency, cost-effectiveness, and ease to implement—in other words, it is a 'low-hanging fruit.'"

Too often, areas are not protected for their ecosystem potential or to mitigate specific threat, but because they are the least contentious to a wide stakeholder group. Protection may be the highest priority, but restoration of degraded habitats (ideally by rewilding) and habitat creation (to replace lost habitat areas) are also "must dos." Well-known societal carbon-offsetting schemes typically focus on terrestrial carbon storage such as large-scale tree planting. This is often ineffective for very many reasons, such as planting at the wrong densities, choosing successional or non-indigenous species, ignorance of soil or sediment structure, and susceptibility of them to extreme conditions or environments.

For carbon offsetting to be effective, the ecosystem capturing and storing the carbon needs to be growing; thus fire, drought, disease, or other loss can reverse carbon-sink offsetting very rapidly and can and must be planned for. Some strong knowledge and understanding already exists to improve efficacy of tackling both crises simultaneously. We need to use climate models better to help predict near-future change to make more of some easy wins of joint climate-nature solutions. Not all of these low-hanging fruit of nature-based solutions are immediately obvious, especially if geographically far from urban centers. We need to get "out of sight" in to mind, as the ocean covers 70 percent of our planet and the polar regions more than a fifth. These areas contain some less-thought-about but climate-valuable carbon- and species-rich habitats.

> "Protection of 30 percent of land and sea by 2030 would be more effective if areas designated are targeted to biodiversity functionality and threat mitigation."

LIFE INTERACTION WITH CLIMATE

Earth's atmosphere and climate has shaped life for hundreds of millions of years. However, life on Earth also influences both atmosphere and climate. This mainly occurs through exchange of the gas CO_2 by primary producers (i.e., autotrophs), such as plants, algae, and some bacteria, to grow in a process termed photosynthesis. We are most familiar with this in terms of trees capturing CO_2 through photosynthesis and storing it in trunks for decades to centuries. Some carbon is also stored in a wider foodweb involving primary consumers (herbivores) and higher (carnivores) and other trophic levels.

On death of such organisms, most of this carbon returns to the carbon cycle (i.e., decomposing or rotting) and, ultimately, has little climate influence. However, small amounts of carbon captured through organisms can be buried in soil and sediment beyond the layer of microbial activity. This carbon leaves the carbon cycle and can provide a highly efficient pathway (flux) from atmosphere to locked-up, long-term underground storage, termed "sequestration." More greenhouse gases, such as CO_2, in the atmosphere contribute to global warming, whereas those that have become sequestered through burial help stem climate change. Over long periods of time, greenhouse gas concentrations and Earth's temperature have varied considerably but usually gradually.

In recent geological time, our planet has alternated between ice ages (about ninety thousand of each one hundred thousand years) and brief warm periods (about ten thousand of each one hundred thousand years), in which there is high ice volume or high sea level, respectively. The cyclicity of this is complex because it is driven by imperfections and wobbles in Earth's orbit around the sun (referred to as Milankovitch cycles). Atmospheric conditions, such as gas composition and temperature, are frozen in time when snow is incorporated into ice and builds up over millions of years on polar land masses, such as Greenland and Antarctica.

Drilling deep into these ice sheets to collect ice cores enables us to analyze conditions in the past. Cutting and dating slices then allows us to investigate the relationship between greenhouse gas composition and planetary temperature. So far, we have explored most of the last million years and found that there is a very strong link between CO_2 concentrations in the atmosphere and the temperature of the planet. CO_2 concentrations in the atmosphere are higher now than at any point within the last million years and are increasing more rapidly. Ice cores show that CO_2 and temperature are rarely static, and some increase would be normal towards the end of an interglacial period, but right now, there is clearly a very substantial human input in this process. This is due to a massive use of so-called fossil fuels (oil and gas) in very many applications, which ultimately all return greenhouse gas to the atmosphere.

> "The polar regions are disproportionately important during warm periods, such as we are experiencing currently."

Drilling into deep mud accumulations in seabeds has also allowed collection of sediment cores. These have shown that sequestration of carbon by life on Earth tends to be highest (and most variable) at the height of glaciations and lowest in the brief warm periods like now. Importantly though, they show how powerful nature can be in "draw down" of carbon from the atmosphere and that the polar regions are disproportionately important during warm periods, such as we are experiencing currently (a point to be returned to later). Since the middle of the last century, humans have massively increased emissions of greenhouse gases to the atmosphere, which caused drastic increases in temperature, ocean acidity, ice loss, weather extremes, and other complex climate effects (such as changes in precipitation). This recent human-mediated climate change can be reduced, halted, and, ultimately, reversed with appropriate action, and nature can be an important part of the solution.

CAN NATURE HELP TACKLE CLIMATE CHANGE?

Life on Earth has long influenced the climate of the planet, mostly via levels of CO_2 drawdown and sequestration, but sometimes profoundly. About fifty million years ago, a large Arctic lake filled with a small, fast-growing fern called *Azolla*, which extracted planet-changing amounts of CO_2 from the atmosphere rapidly. Thick bands of coal evident in Greenland and Canadian islands are testament to how quickly, efficiently, and important carbon drawdown by that fern and life in general can be. This process resulted in Earth's temperature trajectory to cool, which eventually caused the polar regions to freeze and build up into the thick icy landscapes that we have today.

We now have the technology to measure carbon drawdown to storage in many ecosystems of our planet. These habitats can rapidly capture CO_2 and store it in foodwebs and, in the right conditions, bury it effectively. This means nature (biodiversity) can be an extremely efficient regulator of the climate through interaction with the carbon cycle. This helps with adaptation to, and mitigation of, greenhouse gas emission–driven warming. On the terrestrial side, forests can be important as hotspots of biodiversity and in storing CO_2 but are not necessarily good for sequestration. Their power to act on climate stems mainly from increasing storage size and thus net forest size. The process of carbon sequestration is not confined to or even most efficient in the forests most of us see, but widely occurs throughout many global habitats and environments and varies considerably in time and space.

> "Coastal wetland systems are such strong carbon- and species-rich hotspots with so many human health co-benefits that work to protect and enhance them must be high on any priority action list."

Mud underlying coastal wetlands, such as mangrove swamps and seagrass meadows, shows that these habitats are important carbon sinks (accumulate a large mass of carbon) and, more excitingly, particularly efficient at its sequestration (measured as gCO_2 buried per unit area per unit time). Carbon stored in marine habitats is often referred to as "blue carbon." Blue carbon sinks are not confined to mangroves and seagrasses. Other environments, such as salt marshes and macro algae (such as kelp "forests"), are also carbon- and species-rich; however, quantifying their carbon sequestration potential is more difficult because much of the carbon is not stored or buried *in situ* but widely "exported" by tide, the current, and storms. Where this exported carbon eventually ends up varies and can be far from its origins and widely dispersed, for example, at the base of continental slopes, basins, or trenches hundreds of kilometers away.

Mangrove swamps, seagrass meadows, and salt marshes are restricted to warmer waters and occupy a very small area of our planet's surface, which is currently decreasing (due to reclamation for other uses). Pristine, undisturbed blue carbon habitats are much more efficient and substantial in capture to burial of carbon ratio compared with similar recently restored blue carbon habitats. Thus, prioritizing the least impacted of these well-known carbon-sink habitats for protection must be an imperative.

However, even restoration of those that have been degraded have the potential to be much more powerful carbon sinks than most terrestrial tree-planting schemes. Yet, their restoration needs to be carefully planned, executed, and monitored—not least to properly foster local and regional community support and involvement. This is often not the case with the predictable result of a failing restoration scheme. Coastal wetland systems are such strong carbon- and species-rich hotspots with so many human health co-benefits (e.g., coastal protection, food, and tourism) that work to protect and enhance them must be high on any priority action list of "nature-based solutions" (NbS).

To a large extent, actions to limit climate change or nature loss have been carried out through different organizations which have not been taking each other into account until recently. An expert workshop of the Intergovernmental Panels on Climate Change (IPCC) and Biodiversity and Ecosystem Services (IPBES) strongly emphasized (as have many independent scientists) that it is important to tackle both crises together. The report of that panel showed that many of the flagship successes to date are climate-nature solutions that strongly considered human health, equality, consultation, and involvement. Implementation of NbS is often complex, and many hurdles and complications in holistic protection of ecosystems remain.

In political terms, areas of mitigation action are attributable to a given country if they are within its exclusive economic zone and are then termed nationally determined contributions (NDCs). Thus, NDCs can make mitigation effort beyond areas of national jurisdiction (e.g., the high seas) less rewarding to individual countries. For economics, mitigation effort has to add *additional* carbon and remain demonstrable in long-term storage (about one thousand years). This "additionality" (new carbon buried) can clearly be demonstrated in the case of restoration, rewilding, and creation of habitats. The requirement to show/measure new carbon burial does pose problems for protection of existing mature habitats, though, unless they were otherwise destined for destruction (in which case their stock can be considered "new").

Designation and implementation of nature protection is challenging, particularly around coasts and continental shelves with many diverse societal uses. However, there is now a stronger, clearer consensus that synergistic nature solutions anchored in science are key to tackling the current global crisis. This is a particular challenge in the case of the polar regions, which are, in the most part, geographically far from human population, decision-makers, and NDCs.

NATURE-CLIMATE LINKS IN POLAR REGIONS

The physical environment of polar oceans soaks up and stores a disproportionately high amount of CO_2 (not least because cold water can take up more

CO_2 than warm water). This has been and is a very important buffer for the global climate, but it is limited and will decrease with warming. CO_2 in polar oceans can be released (outgassing) and happens at the cost of changing pH (ocean acidification). Until recently, there has been little consideration of the import of the interaction of high-latitude biological carbon pump with climate. Recent scientific work has highlighted that full protection of the marine environment (i.e., no-take zones) lags behind the proportion of protected area on land, with polar seas lagging behind the furthest.

This is somewhat out of kilter with the societal impression of the polar regions as "world park wilderness." This lack of protections is also contrary to findings showing that the seafloor in polar regions soaks up disproportionately high amounts of CO_2 during interglacial periods (such as in the present time) as well as being ignorant of polar biodiversity hotspots. The habitats that led to the coining of the phrase "blue carbon," mangroves, salt marshes and seagrasses, do not occur at very high latitudes, so perhaps it is not surprising that cold waters escaped such attention, yet, they are vital. The biodiversity around the Southern Ocean, for example, is remarkable, not just for the super abundant copepods and krill that feed the largest global populations of whales, seals, and flightless birds, but for its high species numbers, of which the majority do not occur anywhere else and have not even been described yet.

> "Full protection of the marine environment (i.e., no-take zones) lags behind the proportion of protected area on land."

Take the small archipelago of the South Orkney Islands, east from the tip of the West Antarctic Peninsula. Marine life around here is richer and more unique than that around the Galapagos Islands, renowned for its biodiversity. The base of these polar foodwebs is vast microalgal blooms (phytoplankton) and extensive kelp forests, which are super productive throughout the summer season across massive continental shelf areas (6.6 million square kilometers around the Southern Ocean).

Such high-latitude coastlines are often dominated by fjords, which, despite comprising just 0.3 percent of ocean area, store about 11 percent of its carbon. Most compelling of all is that, at least in the Antarctic, marine ice loss is increasing the carbon storage and sequestration potential by life on the seabed and thus works as rare negative (mitigating) feedbacks on climate change.

THE STRUCTURE OF POLAR BLUE CARBON

Polar foodwebs and habitats include large and important carbon sinks, some of which are emergent with climate change. Meaningful protection of such hotspots in the cold would help tackle both climate and nature loss crises and would be a big bang for the buck, so why are we dragging our heels?

The process of biologically transferring carbon from air to sea can be thought of as happening in three stages. Just as with land plants, the process starts with carbon *capture* from the atmosphere (dissolved in surface waters) during photosynthesis. Such algae lose some of this captured carbon through respiration (breathing). The phytoplankton blooms can be huge and span thousands of kilometers and may stretch down more than one hundred meters deep. Maximum productivity, where the micro algae are at their densest, may be as deep as twenty to forty meters. This is deeper than satellites can see (termed Earth observation) and thus most of the volume and geography of global ocean productivity is obscured and likely underestimated. Unlike on land, most marine primary producers (and all in polar seas) are not actually "plants" but instead algae, and they have very brief lives, so that any carbon *storage* stage (2) occurs in other parts of the foodweb, mainly animals.

Less than half of all carbon captured by polar phytoplankton or macro algae is stored in foodwebs, and nearly all the rest is recycled into the water column by bacteria in what is termed the microbial loop (rotting). Vast numbers and biomass of water column (pelagic) zooplankton, such as copepods or krill, which are shrimp-like crustaceans, store much of this carbon. Some of this is eaten by a cascade of higher predators, ultimately ending in seabirds, seals, and whales. Some is exported to deep waters or the seabed (ultimately where sequestration occurs) either by passive sinking of algae directly, by active zooplankton migrations (e.g., from the surface into the deep ocean), mass sinking of fecal pellets from large swarms of animals in the water column, and major seasonal migrations of lipid-rich zooplankton.

These pathways are sometimes termed passive, active, swarm, and lipid flux, respectively. Just like in the water column, there are multiple trophic levels living on polar seabeds (benthos). Many of them occur deeper than where the carbon-capturing algae live. This means benthic foodwebs rely on a "rain" of food from above. In contrast to marine life at lower latitudes, a much higher proportion of known species live on the seabed in polar regions (83 percent of those in the South Orkney archipelago). Many are rare or patchy, and we know very little about them. Many polar benthos are long-lived and may store carbon for centuries. Yet we have to keep in mind that in both pelagic and benthic foodwebs, much of the carbon stored in consumers is also recycled back into the carbon cycle through either respiration or microbial breakdown on death.

The third stage of escaping the carbon cycle is very important to climate and termed "sequestration." A small proportion (about 1 percent) of the original carbon captured through algae and stored in organism (mainly marine animal) bodies escapes microbial breakdown to be buried in sediments. Burial alone does not guarantee escape from the carbon cycle. Bacteria, which are instrumental in carbon breakdown through the microbial loop, are still active in surface sands, muds, and silts.

Organisms living close to the sediment surface can disturb the surface (bioturbation) and irrigate burrows, both of which processes increase oxygenation and thus the potential for microbes to recycle blue carbon from deeper layers of the sediment. Some carbon, though, does eventually become long-term buried, completing a pathway from atmosphere to sequestration to ultimately end up fossilized as rock. Hotspots of the entire carbon pathway or stages of the process occur in particular geographic areas, often in association with high biodiversity, which are of particular interest to protect as synergistically important in both climate and nature crises.

FJORDS, A SPECIAL CASE

Polar blue carbon is most obvious as large macro algal forests in shallow waters. Phytoplankton blooms over the extensive continental shelves and generate much smaller production of blue carbon per unit area, but it may settle out to become sequestered in huge muddy basins spanning millions of square kilometers. However, the superstar amongst very high-latitude blue carbon habitats are fjords (with or without glaciers). These are found throughout the northernmost and southernmost Atlantic Ocean but also in New Zealand and Antarctica. Their cold waters are rich in macronutrients fueling macro algae on their steep-sided walls and dense seasonal phytoplankton blooms in the surface waters. These, in turn, feed rich and abundant populations of consumers (mainly animals). Many fjords have glaciers in their innermost sections, most of which are retreating, and some of which are marine terminating and others land terminating. The glaciers and meltwater that run into these fjords not only supply nutrients for growth of life in them but also contribute a considerable sediment load, leading to rapid sedimentation, accumulation, and burial of carbon.

Thus, although fjords are small in area and volume and only moderate in carbon capture and storage, they are efficient as carbon sinks as well as being rich in biodiversity. Some fjords, such as Kongsfjorden in the Svalbard archipelago (Barents Sea), are well-monitored by international scientific programs running oceanographic and biological time series. Others have barely ever been visited, let alone scientifically mapped.

The physical and biogeochemical conditions in most fjords are likely to be changing, not least in response to higher greenhouse gas concentrations and the warming they generate. In West Antarctica, Greenland, and Canada's arctic islands, many fjords will be subject to seasonal sea ice losses (less of their surface freezing for less of the year) and glacier retreat. We know little about the way these changing conditions interact with each other and with life within fjords. Within the last half century, even remote fjordic organisms have been exposed to an increasing intensity and variety of stressors (e.g., fishing and aquaculture, non-indigenous species, microplastics, and other pollution, as well as climate change). The degree of exposure to these stressors, the combinations of them, and the rates of change varies massively from place to place, though. Many key carbon-rich and biodiversity hotspots are threatened by differing combinations of fishing effort, non-indigenous species spread, warming, pollution, and other factors.

> "Even remote fjordic organisms have been exposed to an increasing intensity and variety of stressors (e.g., fishing and aquaculture, non-indigenous species, microplastics, and other pollution, as well as climate change)."

We suspect that an increase in most of these stressors leads to negative impacts on biodiversity in terms of richness, density, life span, and, thereby, ecosystem services. A variety of biological responses to such disturbances has already been detected in some well-studied fjords, such as at Potter Cove (King George Island, South Shetland Islands, Antarctic Peninsula). Here, Argentinian- and German-led research has shown that glacier-retreat-linked sedimentation has changed the species present in two decades.

The biota on the shallow seabed at Potter Cove has increasingly become dominated by sediment-tolerant species. Studies on further fjords, such as joint UK-Chilean research, has shown how rapidly life can colonize emergent seabed within fjords and how these communities contribute to climate change mitigation. The rich and dense biodiversity of many polar habitats coupled with intense seasonal algal blooms suggest there are probably a wide variety of biological carbon sinks at high latitudes, but fjords are likely to be amongst the most important and easiest to measure, monitor, and designate for protection.

> "Fjords are likely to be amongst the most important and easiest to measure, monitor, and designate for protection."

HOW IS POLAR BLUE CARBON CHANGING, AND WHY DOES IT MATTER?

Some ocean basins and continental shelves have accumulated soft sediments over longer periods of time such that we can measure carbon sequestration in the sediment from the present to before one hundred thousand years ago. This can be undertaken by vertically pushing down a tube into seabed muds and collecting a "core" of mud, of which the youngest part is at the surface and the oldest part at its base. Further studies of the layers in these show that carbon sequestration to marine sediments changes considerably over time, sometimes quite rapidly.

Recently, many land-based carbon stores, such as the Amazon forest, have changed from being net carbon sinks to sources (for example, because of deforestation, erosion, and fire). Similar patterns are evident in many coastal wetlands. However, further offshore, in the open ocean, the fate of carbon from primary productivity to sequestration is less well-measured or understood. Yet, there is

> "Land-based carbon stores, such as the Amazon forest, have changed from being net carbon sinks to sources (because of deforestation, erosion, and fire)."

evidence that pathways to sequestration in the open ocean are under threat. For example, fishing effort hotspots are strongly related to hotspots of pelagic (water column) productivity.

In most temperate and tropical systems where we have information, we can detect losses of carbon- and species-rich area. Yet this is happening despite considerable programs of restoration, rewilding, habitat creation, and designating new protected areas. The exception to these declines might be macro-algae (kelp forests). These seaweeds may only occur in the shallows, but they grow very rapidly and support a diverse and productive associated biodiversity.

Most of the carbon captured and stored by kelp forests is widely dispersed when their blades are fragmented in storms (and can often be seen washed up on beaches). This makes it quite difficult to measure the fate of carbon from various sources, but where macro algae are adjacent to deep waters like continental slopes, up to 11 percent of original production can be sequestered (which is very high). Warming and coastal ice losses are resulting in poleward shifts in macro-algae as new space becomes available and more suitable

> "There is evidence that pathways to sequestration in the open ocean are under threat."

(for example, in temperature) for them. The same opening up (emergence) of new suitable space is also likely to be true of salt marshes and seagrass meadows and is becoming an important new area of interest.

Physical change in polar regions, especially in the Arctic, has been very considerable, with strikingly different variability and predictability in time and space. Ice loss is one of the very obvious changes in the form of thinning, melting, retreat, and, in the case of seasonal ice and snow cover, briefer in duration. The polar regions are, thus, losing ice-associated habitat for many organisms living in it (pagophilic). Loss of white reflective ice (albedo effect) means that heat from the sun is not reflected away anymore but instead absorbed by the now blue ocean.

However, opening up of polar seas also makes them a major exception to catastrophic losses of carbon- and species-rich habitats elsewhere. Less seasonal ice over continental shelves can allow earlier and longer phytoplankton blooms, spread of macro-algae, and growth increases within the foodweb (e.g., seabed animals) feeding on them, thereby storing more carbon. Carbon capture, storage, and likely sequestration in some polar seas seems to be increasing in response to climate change, and thus is working as a negative (mitigating) feedback. These mitigating feedbacks are rare and extremely valuable globally. Although, we really need to understand more about them, such as how big they are, where they are, and why they exist.

ARE SYSTEMS SUPPORTING
NEGATIVE FEEDBACKS UNDER THREAT?

An important distinction should be made here between carbon sinks, mitigating feedbacks on climate change, and biodiversity hotspots; a habitat can be all three, any two, or any one of these. Being any one does not imply also being another. A large forest which shrinks because of climate change is a big carbon sink but a positive (exacerbating) feedback on climate change. A small macro-algae bed which increases in size driven by climate change is a small carbon sink but a negative (mitigating) feedback on climate change. This does not mean that we should not keep protecting or restoring forests. It just means that prioritization for protection could be directed towards the most immediately beneficial habitats.

The largest measured mitigating feedbacks on climate change to date are in the polar regions. One of these is driven by snow loss from hyperboreal taiga forests (high-latitude forests south of tundra), and the others all concern losses of ice over the sea. In both taiga forests and marine ice losses, less snow and ice leads to more light and warmth, promoting growth.

Marine ice loss occurs in the polar regions in three main ways: less seasonal freezing of the sea surface (termed sea ice, or close to the coast as "fast ice"); ice shelf disintegration (where land-based ice sheets meet the sea as vast floating rivers); and glacier retreat (often exposing new fjord habitat). They each alter or expose quite different habitat types.

By far the largest change in area is seasonal sea ice loss. Less sea surface area around the polar regions is freezing in winter and for less of the time. Such losses show complex, "noisy" patterns, with considerable inter-annual variability but also contrasting patterns between adjacent seas and between the polar regions. Much of this happens over deep (greater than one kilometer depth) waters, where we know little about how this influences life there or its functionality. Above the continental shelf, areas that now have less seasonal sea ice cover tend to have longer micro-algal blooms, and macro-algal forests may be extending their range southward. Change is complex though. Long-term environmental research of the American Antarctic station of Palmer shows change in microalgal blooms along onshore–offshore and north–south gradients, in composition and even in cell size of these algae. All of these are very likely to influence foodweb structure and carbon pathways to sequestration. United Kingdom- and German-led work suggests that one such effect is increased provision of food to life on the seabed (benthos), of which some organisms' elements have doubled their growth (and thereby carbon storage) in the last three decades.

This does not necessarily mean that there is more food for animals, as it might just be that food is available for longer, resulting in "more meal times." Cold-blooded animals (ectotherms) can take a few weeks to process a single meal (enzymes work slowly in the cold), so a smaller micro-algal bloom but spread out over twice the period could, for example, double food availability for plankton eaters, assuming that there was always more food "on the table" than could be eaten. In summary, where seasonal sea ice losses occur over continental shelves, they can increase carbon storage in foodwebs through more extensive micro-algal blooms.

In the coastal shallows, though, influences of seasonal sea ice loss can be more complex, with contrasting strong effects. This is particularly evident in changes in iceberg-scouring rate. When the sea surface is frozen in winter, icebergs are mainly "locked in" and unable to move much in wind or current. Decreased seasonal sea ice results in icebergs being more mobile and colliding with the seabed more often. This seabed collision rate is measured annually by scuba divers at UK's Rothera and Argentina's Carlini research stations. Markers on the seabed are surveyed each year to see which have been hit, these are recorded, and the markers are replaced. This way, scientists develop a long-term record of disturbance history on the seabed and how it relates to overlying seasonal sea ice cover and climate. In a high ice-scour year, more than one-third of the shallow seabed can be scoured by icebergs, which can be typically catastrophic to life in those areas.

Investigation of the biodiversity in the same exact areas has shown that seabed biodiversity takes about a decade to recover, and the increased growth (from longer phytoplankton blooms) is more than lost by increased iceberg-mediated mortality. So seasonal sea ice losses may result in loss of carbon storage in the productive, open-coast shallows. However, carbon storage increases in the slightly less productive shelf waters beyond the depth of most iceberg keels. By far, most of the area of polar continental shelves is deep, so overall seasonal sea ice losses result in net gain of stored and, thus, eventually, sequestered carbon.

The next biggest polar processes in terms of both carbon-sink size and mitigating feedbacks both work differently to seasonal sea ice losses but are similar to each other. Ice shelves and glaciers are both the seaward extent/ terminus of land-based ice (ice sheets). Where they disintegrate and retreat, they expose new habitat to light and make it colonizable to primary producers, such as micro-algal blooms, macro-algal forests, and, ultimately, the consumers they support. Ice shelves periodically break up at their seaward edge to form giant icebergs (thousands of square kilometers in area). Life can persist under the ice shelves in dark, food-poor habitats.

> "Growth of animals on the seafloor creates new carbon sinks— which would be the equivalent of major forests being planted."

However, such communities are typically thought to be sparse and very slow growing. Thus, when the ice above them fragments and drifts off, huge coastal embayments can be exposed (such as Larsen A and B on the East Antarctic Peninsula). Earth observation imagery (i.e., satellite-derived) shows brand-new algal blooms in these bays, and rapid colonization and growth of animals on the seafloor creates new carbon sinks—which would be the equivalent of major forests being planted. Although the giant icebergs may also occasionally scour the seabed, thereby destroying benthic biodiversity and recycling carbon that might otherwise have been buried, it is more than offset by the new productive bays generated and what happens in the wake of iceberg drift.

As icebergs melt, they leach out nutrients to fertilize the ocean, thereby generating micro-algal blooms in their wake. Scientists have used satellite imagery to study the new blooms in both the wake of giant icebergs and the new bays opened up where they break out. Such work has allowed estimation of net gains in productivity and carbon capture to sequestration. Overall, a five-thousand-square-kilometer iceberg may thus generate a million tons of carbon accumulated at the seabed. As with most such processes, the net climate change feedback is complicated to calculate as it does not just involve the carbon cycle. Floating ice shelves buttress land-based ice sheets. This means they are acting like a plug and stemming the flow of land-based ice into the ocean. So, when a large section breaks off the floating ice shelf (as a giant iceberg), the flow rate of land-based ice to sea may increase, thereby increasing sea level. Like seasonal sea ice losses, floating ice shelf break-out also changes albedo (reflectivity) and increases heat absorption in the ocean.

The final and smallest of the marine ice-loss carbon sinks and mitigating feedbacks to be considered are fjords. Fjords are, however, powerful in respect to species-rich carbon sinks and mitigating feedbacks, they are just small in global area. Like ice shelf break-out bays, they are emergent habitats that could become very important if undisturbed by human activity. As glaciers retreat along fjords, they expose new habitat for colonization by diverse life. Their cold waters are nutrient-rich for strong foodweb carbon storage, their steep sides can be densely covered in biodiversity, and sediment accumulation on their mud floors is ideal for ultimate burial of algae which sediments out fecal material and animals which die.

However, new productive habitats opening up from ice-shelf loss and glacier retreat, and increased productivity in existing habitats (from seasonal sea ice losses), is likely to attract human resource use. Elsewhere, fishing intensity maps well onto carbon export hotspots, whereas protection maps poorly onto carbon-rich nature in the sea. This is especially the case on polar continental shelves, despite containing the most effective mitigating feedbacks on climate change.

"Mangrove swamps, seagrass meadows, salt marshes, macro-algal stands, flooded forests, swamps, and peatbogs must also be at the top of our nature 'to-do' list."

We know and understand nature around polar coastal and continental shelf areas quite poorly. What we do know is that, in the past and present, they have been important carbon sinks and that, unusually, they work as mitigating feedbacks against climate change. Globally, we need to be much more synergistic in our approach to combating nature loss and climate change and put a much higher value on nature. It is vital that we rapidly identify carbon- and species-rich hotspots and protect those least damaged so we can halt the hemorrhaging of global wilderness. Fjords and many polar and subpolar coastal habitats offer some low-hanging fruit of high nature-climate reward for well-targeted, scientifically unpinned protection. Mature mangrove swamps, seagrass meadows, salt marshes, macro-algal stands, flooded forests, swamps, and peatbogs must also be at the top of our nature "to-do" list.

What do we need to do to become more effective in utilizing NbS? Much stronger international collaboration at the policy (rather than just science) level is needed together with less compromise in where is protected and why. Protection needs to be better thought out so it mitigates against specific threats and needs to be well coordinated so key systems do not fall through the gaps (e.g., areas beyond national jurisdiction just because they don't comprise nationally determined contributions for any specific country).

A strong protection framework can then be bolstered with most-promising rewilding and restoration projects. There is likely to also be a strong case in some areas for assisted carbon sinks, for example, by creating and sinking macro-algal rafts, but this needs careful examination and understanding so that it doesn't become counterproductive like much afforestation. We need a properly joined-up, ecosystem-based approach to halting biodiversity loss and meaningful action to reduce climate change, and evidence shows that we must consider nature throughout that conversation.

Chapter Fifteen

Arrogance, Carelessness, and Humility

BF Nagy

Scientists and other experts treat biodiversity loss, species extinction, greenhouse gas emissions, air pollution by particulates, water pollution, and ocean warming as distinctly separate because they understand them in terms of the differing measures needed to mitigate each of them. But to me, they are all caused by one thing: our arrogance as a species. Our collapsing world is not the fault of China, capitalism, six specific billionaires, corrupt politicians, or even the truly evil oil companies. It's caused by all of these, especially by our general lack of humility and the belief that we have a right to keep abusing our only home and that it will withstand our abuse indefinitely. But, as we know now, it won't. We must rapidly change.

WITH MORE THAN EIGHT BILLION OF US, WE'RE DOING DAMAGE EVERYWHERE

For much of my life, I was careless about protecting our planet. We need to accept that none of us can keep thinking we're exempt, or are excused from helping solve the crisis, or have time to get around to it later. When I asked marine ecologist Dr. David Barnes about life-changing experiences, he described a research expedition some years ago when his team was studying a remote, largely unknown area in the South Atlantic Ocean where wave swells might be thirty to forty feet high, water and air temperatures between 40° and 50°F, and wind at perhaps fifty miles per hour. They were thousands of miles from civilization using cameras to study mountains on the frigid sea floor about twenty-three thousand feet below the surface. South Atlantic seamounts are often formed by volcanic activity and are critical habitat for some ocean life.

The team was suddenly caught by surprise and could scarcely believe what they were seeing. "We were trailing our live-view drop camera," said Barnes. "We could see the extreme contrast of lasting damage where a fishery trawl had been dragged across (the seamount peak) years before and, just meters away, undamaged centuries-old, rich, cold-coral reef. Then we saw some discarded plastic on the seabed! It reminded us that even the furthest places from busy human activities face complex stresses of our origin."

LOCAL GOVERNMENTS

I had dinner several years ago with Ralph Torrie, a brilliant climate-modeling innovator who worked for the United Nations and inspired some of the ideas in this book. He said we are all responsible for working together to solve the climate crisis, but that some people were working in roles which could have a disproportionately profound impact. One of the examples he gave was the largely unheralded urban planner. I reached the same conclusion after working in government in the City of Toronto. Urban planners create reports, recommendations, and development policy. They approve the details of all key developments. They shape cities, and experts say that one day soon three-quarters of humanity will live in cities.

As mentioned, cities and other local governments are nimble, connected to the people, progressive, green, and action-oriented.

BUSINESS

We've learned that businesses sometimes fear early adoption of green programs and seek mandated standards and level playing fields. We've also learned that once the regulatory framework is in place businesses can make a huge impact on our climate-crisis imperative.

I love visiting Portland, which is called the City of Roses and celebrates these iconic flowers with a dedicated festival each year. And like many parts of the West Coast, I think of Oregon as a big boreal forest with smart people hiding behind every oak, hemlock, and fir tree (along with pelicans, woodpeckers, and beavers). Kate Gaertner is one of these, and she looks, sounds, and talks like a savvy business executive. But she's also different, and I wish more business leaders were like her, with her penchant for action and measurable results, her understanding of sustainable economics, and her modern approaches to business investment and opportunity. To Gaertner, going green is sound strategy. Rather than threaten financial results, sustainability guarantees them.

Humility: "As a society, we have adopted this notion that what we want is what we need and that there are no consequences to what we consume," says Gaertner. "If we can consume and manufacture less and do more with what we have, we can decrease the burden of consuming resources, allow natural resources to regenerate, reduce if not eliminate waste, and live within the boundaries of the Earth system." Through OMALA, her sustainable fibers activewear company, she spent a great deal of time engaging in dialogue with customers about the value of wearing sustainable textiles. It motivated her to return for her second master's in Sustainable Management. "I wanted to propagate the concepts that undergird circular economy within businesses and industry. Today—every day—I work to challenge ingrained assumptions, build financial models to prove value, and implement circular concepts to show proof of concept that living and operating sustainably is our destiny."

Chapter Sixteen

Circular Food Systems
Feeding the Urban World
Kate Gaertner

Kate Gaertner of TripleWin Advisory in Portland, Oregon, holds an MBA from the Wharton School at the University of Pennsylvania, an MSc in Sustainable Management, and is the CEO of TripleWin, which serves clients such as the Linux Foundation and StockX. She is an expert on progressive corporations, global food systems, circular economy, and greening company supply chains. Gaertner has written for *Forbes*, *Fast Company*, and *GreenBiz*, and is the author of the award-winning *Planting a Seed: 3 Simple Steps to Sustainable Living*. One of her first business ventures was an activewear company that designed and manufactured clothing made from sustainable fibers.

We live in a world of intensifying climate change, growing human populations, and excessive food waste: a triad of trends feeding off each other in consequential and detrimental ways. Our agricultural practices contribute substantially to the climate crisis. Project Drawdown[1] estimates that nearly 25 percent of all atmospheric greenhouse gas emissions are due to our industrial agricultural system, global food chain, and land conversion for cultivating crops and raising livestock. The climate crisis jeopardizes crop productivity globally, challenging our ability to feed the world's populations. Eliminating most loss and waste in our food system can close the gap on calories produced but not consumed by 12 percent,[2] helping to feed the additional 2.5 billion people living on the planet by 2100. Rethinking and localizing our food systems can bolster resiliency, reverse climate change, and feed more people well.

INTENSIFYING CLIMATE CHANGE

The third agricultural revolution made the global food system more efficient and productive but heavily reliant on carbon-emitting fossil fuels. It also ushered in a monoculture farming mentality that globalized cultivation of a small number of high-demand crops—soybeans, wheat, rice, and corn—a profit boon for large food conglomerates but a significant weakening of regional food sovereignty and cultural food heritage.

The focused intensity on growing the "big five"[3] commodity crops to feed the world's population is also a major perpetrator of carbon emissions into the atmosphere. Rice cultivation alone contributes a staggering 1.5 percent of the globe's greenhouse gasses and emits 12 percent of global methane emissions.[4] Wheat crops are highly dependent on synthetic fertilizers and pesticides for their productivity and require diesel-fueled farm equipment to be harvested. Soy's climate change contribution is mostly due to land conversion to cultivate crops; removing natural carbon sinks in favor of arable land to be used for intensive—but not more productive—farming. Our global agricultural system necessarily comes with increased transportation and distribution miles traveled to get food products to their final market destinations, a secondary yet substantial driver of carbon emissions.

The food system we have in place today, one we have nurtured for a half-century and that produces some 43 percent of the world's calorie supply,[5] is itself in jeopardy due to the growing climate crisis. Rising temperatures and prolonged droughts are withering crops; frequent floods are destroying them.[6] The reliance on intensive farming practices such as tilling and heavy fertilizer and pesticide use robs the soils of their productivity, jeopardizing yields necessary to continue feeding rising populations.

GROWING POPULATIONS

Human populations will grow significantly larger within this century with an expected 9.8 billion people to feed by 2030 and nearly 10.5 billion by the end of the century. Those populations will increasingly be living in urban areas across the globe, from 55 percent today to 63 percent by 2050. Current urbanization in Europe and North America has far exceeded those developing region estimates, putting the number of city-dwellers at 74 and 82 percent respectively.[7]

This human migration to higher-density living away from greenfields and farmlands to concrete jungles has cascading repercussions, from greater food insecurity and lack of access to fresh, nutritious food, to a collective forgetting of where food and physical health originate, to growing waste streams that feed the climate crisis and pollute human communities and their naturescapes. Our globalized industrial food system combined with concentrated urban living makes for a fragile and disconnected system for feeding human populations well.

FOOD INSECURITY

The global pandemic of 2020 proved just how dependent the world's food system was on globalized trade and transportation systems. Most modern cities are "almost exclusively" reliant on imports for their citizens' most basic resources and daily food consumption needs.[8] This import model translates into more expensive and less fresh, nutritious food options. So, too, cities are often "food deserts" that inhibit the ability of residents to easily access nature-grown food. The US Department of Agriculture (USDA) reported that 39.4 million US citizens living in cities had low access to affordable and nutritious food. In the European Union, 8.6 percent (38.4 million) of the population were unable to afford daily nutritious, protein-rich meals.[9] Nearly double the number (68.1 million) of individuals living in urban areas in sub-Saharan Africa were at risk of acute food insecurity in 2020.[10] The time, transportation, and personal logistics costs of eating well in urban areas become reinforcing barriers to most.

As people move away from tending the land for sustenance to laboring in factories for work and office buildings for wages, their relationship to food changes. The knowledge of planting seeds to be nurtured and harvested into delicious food that is shared in recognition of nature's bounty is quickly forgotten. City food becomes something packaged, purchased, consumed, and discarded: a commodity product with minimal perceived value. This disconnection to how and where food is grown devalues the work, effort, and natural resources (e.g., water, soil, sun, and seeds) that go into feeding the world's populations. Because of their nature-poor existence, urban dwellers can quickly lose their sense of awe and wonder of where food comes from and how delicious, sensuous fruits and vegetables feed the soul and body in simple and effective ways.

UNSUSTAINABLE FOOD WASTE

Waste is a direct by-product of the disconnection between urban dwellers and their food sources. In high-income countries, 40 percent of food waste occurs at the retail and consumer levels.[11] At retail, food is discarded mainly due to consumers' high appearance standards. At the individual level, consumers tend to overbuy or fail to eat all the food they order at restaurants or prepare at home. In North America, 1,520 food calories[12] are lost or wasted per person per day. That is more than double the calories thrown away in Europe and industrialized Asia. All told, some 1.3 billion tons of food is lost or one trillion US dollars are thrown away each year.[13]

> **"Reconnecting people to the food they eat is likely to lead to decreased waste."**
> —Matei Georgescu, Associate Professor of Geographical Sciences and Urban Planning at Arizona State University

All this waste—wasted human resources and productivity, natural resources, and edible food—needs to be rethought. Food waste is responsible for nearly 8 percent of global emissions.[14] With climate change continuing to rise unabated and where 2.3 billion people globally are food insecure, recovering food calories produced is an urgent challenge to solve.

BUILDING RESILIENCY

Given the reality that a large majority of humanity will be living, working, eating, and seeking to thrive in urban environments by 2050, our responsibility as community builders and policymakers is to create sustainable food systems that meet the needs of diverse populations in equitable, affordable, and healthful ways. With climate change and its attendant extreme weather events; the growing risk of morbidity and disease vectors northward; and human strife bringing uncertainty and disruption to global food supply chains, building long-term sustainable and resilient urban food systems has become a priority.

Food circularity—A holistic approach to building food resilience is through the understanding and application of circular food systems within urban environments. Concepts such as zero waste, industrial symbiosis (or co-location), eco-efficiency, shared resources, and regenerative farming practices help to foster food-resilient cities and their communities while improving resource security, building vibrant local economies, and improving public health outcomes.[15] We live in an increasingly chaotic, uncertain, and populous world. Adapting to this evolving environmental landscape requires a fresh perspective that marries old sustainable practices with technological innovations and community-based solutions for feeding the world.

A 2018 study of urban agriculture made the argument that city farming could advance not just food security and waste reductions but also support a lighter environmental footprint by tamping down urban areas' heat island effect through increased vegetation cover; removing carbon dioxide and methane emissions through landfill diversion; and boosting species biodiversity by offering natural habitat for key pollinators.[16] The Earth-system benefits of our agricultural system can be amplified by making urban food systems circular. The Ellen MacArthur Foundation reports that a large-scale circular food economy would reduce greenhouse gas emissions by 4.3 billion metric tons of carbon dioxide per year, or remove some 12 percent of total carbon emissions from the atmosphere,[17] a key solution in helping to reverse climate change.

What does food circularity entail? What do examples of its implementation look like? What follows is a discussion of the key tenets of an urban circular food economy and examples of the concepts in practice.

INDUSTRIAL SYMBIOSIS (OR CO-LOCATION)

Industrial symbiosis or the concept of circular co-location is the idea of having companies situate in close proximity to each other and cooperate to funnel waste or by-products of one entity into key resource inputs for others' processes. The implementation of industrial symbiosis helps to create a closed-loop system of operation where waste can serve multiple purposes such as generating energy (biogas), upcycling into new products, and becoming a beneficial feedstock for another process. The key advantages of co-location are waste elimination, material efficiencies through upcycling and downcycling opportunities, and input cost savings spread across multiple entities.

Industrial symbiosis can be efficiently deployed in high-density urban settings where economies are complex and offer a range of small-to-large companies across a diverse industry base. Circular food hubs can be established in redeveloped brownfields or within abandoned warehouses that support social-impact food upstarts and new food technologies to form a hive of innovation that brings economic vibrancy and entrepreneurial creativity to a city.

> "Last-mile distribution of food products can eschew long-haul truck transportation in favor of intra-city delivery mechanisms such as e-scooters, ebikes, and electric tuk-tuks, reducing systemwide distribution costs and emissions footprints."

Urban systems also reduce food miles traveled, given their embedded location within cities and proximity to high-density dwellers. Last-mile distribution of food products can eschew long-haul truck transportation in favor of intra-city delivery mechanisms such as e-scooters, ebikes, and electric tuk-tuks, substantially reducing systemwide distribution costs and emissions footprints.

In Chicago, Bubbly Dynamics, a social enterprise that supports urban industrial development, has converted a former meatpacking facility into a hub for twenty symbiotic local food businesses that "feed" off each other in dynamic, beneficial, and circular ways. On the property is a 93,500-square-foot anaerobic digester, which turns organic waste into compost to enrich soils; biogas used to electrify and heat the facility; and a nutrient-rich liquid fertilizer to grow algae, a food product ingredient.[18]

Housed within the warehouse innovation hub is a kombucha company where carbon dioxide from the yeast-fermenting process is used to stimulate the growth of plants and algae. Wastewater from an ice-cream manufacturer is channeled into man-made wetlands located in the lobby that supports fishes and turtles as well as irrigates a vertical living plant wall. And Back of the Yards Algae Sciences grows algae supported by food compost to create alternative meat products.

> **"A hub for twenty symbiotic local food businesses that 'feed' off each other in dynamic, beneficial, and circular ways."**

In St. Louis, Missouri, Anheuser-Busch (AB) InBev has invested in Evergrain, an upcycling startup that takes spent barley grain from AB InBev's beer-brewing process to produce protein isolate. AB InBev realized a decade ago that its spent barley grain was being wasted. It was either being sent to landfills or used in animal feed. Through research, the company found that 30 percent of spent grain material consisted of highly nutritious protein that, if isolated, could be a valuable plant-based input to food and beverage products to help feed the world.

Although Evergrain is a separate company, it is co-located on AB InBev's manufacturing campus. AB InBev spent $100 million to refurbish a 1905 building on its property that had gone unused for thirty years. AB InBev built a pipe from its brewing factory directly to the Evergrain facility so that all "barley saved grains" flow immediately and directly to the upcycling facility to be managed and distilled. The companies' mutual proximity to each other means "Evergrain can have direct access to the thousands of tons of 'spent' grain that AB InBev produces, reducing the company's environmental impact."[19]

In Dobbs Ferry, New York, along the Hudson River, Spare Foods Company uses whey, a natural by-product of cheese and yogurt production and the remnants of milk after it has been acidified, curdled, and strained, to produce an effervescent fermented beverage similar to kombucha but more nutritious and protein-rich. The company purchases all of its whey from a symbiotic yogurt manufacturer, the White Moustache Yogurt Company located in Brooklyn, New York. This relationship serves multiple purposes: The White Moustache sells its "waste" to a company that uses it to make consumer beverage products; Spare Foods has a reduced cost of materials and a steady supplier of those inputs in close proximity to its beverage manufacturing plant; consumers benefit from a product on the market that is high in protein, vitamins, and electrolytes; and there is a reduced environmental footprint for both companies.

ECO-EFFICIENCY

Eco-efficiency is an implementation strategy that focuses on doing more (e.g., productivity, output, etc.) with fewer inputs (e.g., energy, materials, water, time, etc.).[20] Eco-efficiency is a circular concept of seeking to use fewer environmental resources, generate less waste, and decrease environmental impact (or pollution) while creating more of what is needed.

Within an urban environment, two eco-efficiency strategies can be employed: utilizing existing built infrastructure to plant rooftop gardens and vertical farms and implementing hydroponic greenhouses that require less space and natural resources yet can produce diversified and nutritious year-around crops. Especially in northern latitude–based cities, rooftop hydroponic greenhouses optimize typically short growing seasons.[21]

Hydroponic systems at scale support urban food security and species biodiversity, and optimize natural resource use by significantly reducing the environmental footprint of traditional crop cultivation. The hydroponic water system reuses and recycles itself, typically from captured rainwater. Crop soils are maximized, as is space management. Because the hydroponic system is self-contained and vertically tiered, soil depletion and typical nitrogen runoff from synthetic fertilizers and pesticides are limited. Cities can produce fresh, diverse, and local food all year long with "virtually no waste."[22]

In Massachusetts, on top of the Boston Medical Center, visitors will find a highly productive rooftop farm that cultivates more than twenty-five crops,[23] supplying the hospital's patients with fresh, local produce while extending food security to low-income patients through its Preventive Food Pantry, a free food outlet. "Most urban environments are food deserts. It's hard to get locally grown food and I think it's something that we owe to our patients and our community," says David Maffeo, the hospital's senior director of support services.[24]

> "In Massachusetts, on top of the Boston Medical Center, visitors will find a highly productive rooftop farm that cultivates more than twenty-five crops, supplying the hospital's patients with fresh, local produce."

In Brussels, Belgium, a rooftop garden is cultivating over sixty species of plants on the roof of a supermarket. The mission of the Lagum Project, as it is called, is to get more of the municipal residents of Ixelles to cultivate gardens themselves. The fruits and vegetables grown are distributed and used by Refresh, a social-impact restaurant serving meals to vulnerable community residents. The multi-year pilot project is financed by the European Union in coordination with the municipality and the Agroecology lab of the Brussels Bioengineering School.[25]

A rooftop in Tel Aviv, Israel, has scaled food production to feed its citizens. On the top of Israel's oldest shopping mall, Dizengoff Center, located in the heart of the city, the "Green in the City" farm produces ten thousand heads of leafy greens a month year-around as well as seventeen different kinds of vegetables and herbs inside two no-soil hydroponic greenhouses that total 795 meters of growing space.[26]

The rooftop project was launched in 2015 by LivinGreen, a hydroponics company, and Dizengoff Center's sustainability department to meet the challenge of feeding urbanized populations affordably and healthfully. The hydroponic farm uses a deepwater culture foam raft system that allows vegetables to be grown twice as fast as traditional methods and with far less product spoilage and water usage and without the use of pesticides and synthetic fertilizers.[27] The fresh produce is sold to both restaurants and households, with deliveries made by bicycle within the city itself.

ZERO WASTE

Waste is a byproduct of linear systems humans have created and perpetuated. Food waste is particularly troublesome in urban areas. In cities such as Beijing and Shanghai, food waste comprises on average 60 percent of municipal solid waste.[28] In the United States, more food is thrown away than anywhere else in the world, making up 22 percent of the country's municipal solid waste at a market value of $218 billion, or the equivalent of 130 billion meals.[29]

> "In the United States, more food is thrown away than anywhere else, making up 22 percent of the country's municipal solid waste at a market value of $218 billion."

People who are disconnected from where their food is grown and reliant on highly processed and plastic-packaged food products end up throwing more food stuff away along with the packaging that encases it. The World Resources Institute pins the high level of food waste in cities on three interrelated consumer behaviors: the low cost of food relative to disposable income; the very high standards consumers have for how their food should look; and a lack of appreciation for how resource-intensive food production is.[30]

Circular food systems seek to reduce, if not altogether eliminate, unnecessary waste, particularly waste that holds innate value. Instead of wasted food being dumped into landfills that off-gas three billion tons of greenhouse gases into the atmosphere each year, that edible food can be diverted, repurposed, and converted into inputs that help mitigate climate change and provide nourishment to those in need. Urban environments are also rife with opportunities to move edible food in efficient ways into the hands of its most vulnerable populations and for cities to implement policies to support the beneficial conversion of food waste at the residential and commercial levels.

FOOD DIVERSION

The problem is not that there are not enough crops cultivated and food produced to feed the world; it is that too much of what is cultivated is lost or wasted along far-flung, globalized supply chains. In rich countries, 40 percent of the food for human consumption is wasted at the very end of the food value chain: at supermarkets, restaurants, and in homes. In developing countries, that number is flipped, with food loss occurring at the very beginning stages of food production, when food is not harvested and rots on farmlands or is ruined by pests or mold in inadequate storage facilities.

Better food management addresses two global challenges: hunger and methane emissions from landfills, which scientists note are of particular concern in the next ten to fifteen years. In industrialized countries, a traditional mechanism for distributing perfectly edible food, whether perishable or packaged, is to establish urban food banks to funnel excess food supplies and uneaten meals to centralized locations that support feeding the hungry, no matter what their circumstance. Sometimes, however, the problem is ensuring that food products *make* it to the food banks in order to get distributed to those who need them most.

One closing-the-loop business found in the United States is Move For Hunger, which seeks to eradicate food waste during transition times—when people move out of their homes and relocate elsewhere. Its national network of transportation trucks receives residential left-behind food and distributes it to food banks that can move the products into the hands of the hungry.

For food retailers—bakeries, stores, groceries, and restaurants—there is a growing global industry that leverages technology accessed by people's smartphones to connect customers with sellers of unsold, surplus food. One of the first movers using technology to close the food chain is the Danish company Too Good to Go, which utilizes its phone app to allow individuals to identify in their local area where edible food is in jeopardy of being pulled off shelves or thrown away. Too Good to Go allows retailers to alert consumers of significantly marked-down, near-expired, and unsold food that is available for immediate purchase.

Different business models are needed to combat food loss in the developing world. Here is where the volume of material throughput needs to be managed more effectively to ensure that food is harvested and food product is efficiently distributed to individuals who need and/or can use it. Gleaning and food-recovery organizations are effective in ensuring that mature vegetables and fruits get off the farm (or urban gardens) and into human hands. Retaaza, an upstart company in Atlanta, Georgia, could serve as a replicable model for expediting local food value chains in order to get food to communities, build stronger rural-urban bonds, boost local economies, and ensure food security of the most vulnerable. Retaaza seeks to ensure that cultivated food within a region is distributed across a spectrum of partners and organizations, including businesses, food banks, and retailers, and gets distributed fairly and consumed, all the while paying fair prices to farmers.

REPURPOSED FOOD

Food waste will always be a reality. But that wasted food can be reduced significantly by the many creative, innovative, and sometimes obvious ways in which it can be reused, upcycled, and downcycled for human and non-human (think: pets) consumption. An old idea that is being implemented in large-scale food production systems like the Oregon Convention Center's restaurant and catering service operation is to give a "second life" to uneaten foods. The catering staff seeks ways to reuse food a second time to eliminate wasteful practices such as freezing fresh fruit to make baked muffins and smoothies; taking stale croissants and breads and repurposing them into bread puddings; and saving carrot tops to use in stews and for making vegetable broths.

Miffy's Foods, located in West Linn, Oregon, works with grocery retailers to rescue overripe fruits from landfill for use in her baking mixes. Another Oregon-based company, Portland Pet Food, recovers high-in-fiber-and-protein spent beer grains and upcycles the material into delicious and nutritious pet treats. Lastly, companies like Imperfect Foods based in California work with farmers and retailers to distribute second- and third-tier produce direct to consumers. The company takes fresh fruit and vegetables that have undesirable cosmetic quirks and irregular sizes that are often rejected at retail by consumers and ensures they are eaten rather than landfilled.

FOOD CONVERSION

Another way to divert food waste from landfills and to substantially reduce greenhouse gas emissions attributable to urban municipalities is to offer citywide composting of both residential and commercial food waste. The compost can be sold back to farmers both rural and urban to regenerate their soils, making them more productive and resilient to rising temperatures and droughts. Urban municipalities can invest in large-scale anaerobic digestion systems to convert food and other organic matter (such as paper and cardboard, yard trimmings, grass clippings, and leaves and wood) from waste to energy, creating renewable power that supports the greening of a city's grid.

NATURAL, SUSTAINABLE, AND REGENERATIVE INPUTS

A circular food system is necessarily composed of inputs that are natural, sustainable, and renewable by nature. Regenerative farming practices are central to a resilient urban food system. The concepts of using no fertilizers, herbicides, and pesticides on crops; biocontrols or "good bugs" such as parasitic wasps, ladybugs, and predatory mites to keep pests under control; no-till plots; and a diverse mix of weed-suppression cover crops to cultivate food are practices showing promise in urban farming.

Organic farming allows farmers to use resources more efficiently to produce more healthful and plentiful fruits and vegetables. By incorporating cover crops, mulches, and compost that improve soil quality, conserve water, and build up nutrients in soils, the water burden of urban agriculture systems can be reduced substantially and synthetic chemical use altogether.[31]

LIFE-ENHANCING FOOD SYSTEM

The crux of city living is the disconnection between people and what sustains them. Bringing food cultivation closer to home—nature to cityscapes—enriches the human experience in meaningful ways. Urban farming, community gardens, and edible landscapes foster community and the sharing of resources between neighbors and among a diverse spectrum of people. City residents can see seeds growing into something plentiful, edible, nutritious, and delicious. They can experience first-hand the life cycle of crops and the critical pollinators and insects that support nature's fecundity.

Practically, urban food systems support food security of city dwellers by providing local, seasonally-grown fresh fruits and vegetables that can be typically hard to find and expensive and offsetting the cost of grocery bills.

"The United Nations Food and Agriculture Organization (FAO) reports that 80 percent of the world's food will be consumed by cities in 2050."

These urban food places often become convenor hubs where rural farmers drop off their produce to supplement and enhance what is grown in the city.[32] In this way, a greater connection is forged between urban and rural communities while supporting economically the broader community of food actors. Yolanda Gonzalez, an urban agriculture specialist with the Cornell Cooperative Extension in upstate New York, equates urban farming with sovereign food production: the right of peoples to healthy and culturally appropriate food produced from ecologically sound and sustainable methods. She says, "For many people, growing food is an important step toward exercising their rights."[33]

We know that to feed nine billion people by mid-century, crop calories must shift from feeding livestock for meat production to going directly into human stomachs. Urban farming emphasizes a plant-rich diet that is *in situ*, which offers a double benefit: fostering a healthy, well-balanced, and globally sustainable diet with an environmentally light footprint.

Circular food systems provide a holistic approach for re-imagining an agricultural value chain that is currently globalized, industrialized, and susceptible to decreasing productivity, wastefulness, and growing carbon contributions to climate change. Urban systems are local and community-based and foster both food equity and sovereignty among citizen residents by keeping who produces, processes, procures, and consumes healthy foods within a virtuous closed-loop. If city and citizenry resilience is a priority, food circularity delivers both by prioritizing regenerative production, favoring reuse and sharing practices, reducing resource inputs and environmental pollution, and ensuring resource recovery for future uses.[34]

Planners and policymakers will need to contend with various challenges in designing local circular food systems, dependent on the existing urban policies, ordinances, and zoning laws that rule current city decision-making and action. Specific challenges will likely include finding public land to designate for urban farming, accessing irrigation water, ensuring healthy soils for productive crops, earmarking funding, and overcoming prohibitive policies that prevent citizens from cultivating food.[35]

From a policy action standpoint, cities have four main levers at their disposal to progress a circular food system: 1) regulatory and planning, 2) economic, 3) cooperation, and 4) education and knowledge-building.[36] Although not exhaustive, what follows are suggested actions and policy instruments that can be implemented to support a holistic, closed-loop food system within urban environments.

REGULATORY AND PLANNING

City planners can formalize curbside food and yard trimmings compost collection at both the residential and commercial levels while investing in wastewater treatment facilities that provide the ability to convert bio-sludge into three valuable components: biogas fuel to be used in public transportation systems and municipal buildings; compost to sell to urban farmers, farms, and residential gardens; and liquid fertilizer to support the regeneration of nutrient-rich crop soils. Critically, municipalities need to rezone city-owned and public lands for urban farming while allowing for sustainable crop production on commercial and industrial buildings. Formally banning food waste to landfills allows for new and innovative approaches for keeping the local food system circular.

ECONOMIC

Promote public-private partnerships to incubate, pilot, and scale circular food production hubs across the city. Support food cooperatives and sustainable, local food production. Create or promote food donation platforms such as Too Good to Go that move perishable food faster at retail and help eliminate food to landfill and hunger among the food insecure. Forge alliances with large municipal healthcare systems[37] to offer primarily plant-based meals to inpatients. Develop a seed bank to support the local biodiversity of plants and food products.

COOPERATION

Allow access to unused urban spaces (e.g., vacant lots, buildings, rooftops, and other unused infrastructure) for residents to plant edible landscapes. Promote local, regional, and native seed use among city gardeners and farmers. Partner with universities and academics to commission research into local food waste–prevention solutions.

EDUCATION AND KNOWLEDGE-BUILDING

Build citywide awareness among residents on how to plant, use, and cook with native seeds through a shared recipes database. Promote the sourcing of fresh vegetables and fruits from local community-supported agriculture (CSAs) programs. Develop guidelines for how to plant sustainably on public lands. Inform city dwellers of gleaner organizations that support the distribution of fresh fruits and vegetables locally. Offer training on the benefits to health and the Earth of plant-rich diets. Educate citizens of the value of "ugly food." Emphasize local distribution of food via "clean" last-mile transportation systems.

Growing more food in and near urban areas makes clear sense. FAO reports that "80 percent of the world's food will be consumed by cities in 2050." Producing food locally and sustainably with an emphasis on plant-based options in close proximity to the world's human populations can go a long way toward solving the twin challenges faced by humanity: climate change and world hunger.

Chapter Seventeen

Water

BF Nagy

Dennis Hayes rose to prominence in 1970 as the coordinator for the very first Earth Day. He expanded Earth Day to more than 180 nations. Most recently, he operates the International Living Future Institute (ILFI), which administers programs such as the Living Building Challenge, a certification that's even tougher than Passive House.

On Earth Day 2013, he opened the Bullitt Center, the ILFI headquarters in Seattle, one of the greenest buildings anywhere. It's particularly notable for rainwater, greywater, and blackwater systems. I visited in 2019 and was amazed. More than a decade after it was built, it remains one of the best examples of urban water management that I've seen. And it's also energy efficient and more beautiful, comfortable, and inspiring than most office buildings.[1]

A few years ago, Hayes told me: "We have a four- to six-week supply of drinkable rainwater at any one time. We had to work hard to get local approval and create a compliance path for other buildings. . . . One of the best resources we have is the seawater that Mother Nature deposits on our roof. . . . There are many places like Seattle. It's getting hotter, and we depend on two reservoirs with declining relative capacity."

Clean freshwater is not always available where and when we need it. Half of the world's freshwater can be found in just six countries. More than a billion people live without enough safe, clean water. And this grows worse daily. In 2023, Jeff Goodell released a book worth reading about our superheated planet called *The Heat Will Kill You First*; however, in the chaos of our apocalyptic future, our families may also die from intense storms, fires, disease pandemics, or lack of food and safe drinking water. All are preventable, and we have more solutions to prevent water pollution than ever before.

The pressure on water resources across many areas in North America affects most of our population. There is a resurgence of interest in technologies that reduce contamination of aquifers and local water bodies. Rules around systems for managing stormwater and sewage are becoming more demanding and are not always followed. In addition to irresponsible profiteers, municipalities with inadequate infrastructure upgrade budgets and contribute to the problem. A main theme in this book is that we react too slowly to technology that emerges as obsolete, like the lead drinking-water pipes in Flint, Michigan, and in thousands of other cities in North America.

By the early 2020s, about three thousand places in the United States were revealed as having higher lead contamination in drinking water than Flint (the infamous example), and more than seven hundred urban communities had crumbling sewer infrastructure that needed upgrading. The Environmental Protection Agency (EPA) says about seventy-five thousand overflows were sending untreated sewage into American waterways every year. The Government Accountability Office reported in 2013 that forty of fifty state water managers were dealing with shortages, or they expected to be in the next five years.[2] To pay for infrastructure improvements, cities have been increasing water rates and other fees, and in 2021, the US Congress chipped in with the Infrastructure Investment and Jobs Act.

A human can survive for months without food but only for three to seven days without drinking water. It's a matter of life and death and a civil matter. If you don't have water, you may have to pick up your family and move somewhere else. Those of us who pollute the most water will eventually be sued or otherwise punished by those of us who pollute the least water. This is true for water and for all of the ways we continue to stress the planet. There is no more time for the absurd free polluting, partisan politics, and delays of the past. The technologies and rules that protect water are proven and uncomplicated, but if public infrastructure is not maintained over time or if big users try to skirt their responsibilities for long periods, water protection can quickly become incredibly costly.

ALGAE BLOOMS

Another discouraging development is the growing incidence of lakes that are off-limits to children and pets for swimming in summer. Data from the Environmental Working Group (EWG) indicates that algae bloom reports grew from about two hundred in 2016 to more than five hundred in 2019.[3]

According to the EPA, too much nitrogen and phosphorus have entered the environment, and nutrient pollution has impacted streams, rivers, lakes, bays, and coastal waters for several decades, resulting in environmental and human health issues.[4] Algae blooms severely reduce oxygen in water, leading to illness or death of fish and other marine life. Some are harmful to humans and pets because they produce elevated toxins and bacterial growth. The EPA says the primary sources of excess nitrogen and phosphorus are improperly managed animal waste runoff, runoff of chemical fertilizers that are not fully utilized by plants, urban stormwater and wastewater overflows, citizens and businesses overusing fertilizers, and detergents, pet waste, and elevated nitrogen in the air from fossil fuels.[5] Obviously, we all have a role to play in preserving freshwater for drinking, swimming, and restoring health to our waterways.

CLEAN WATER SYSTEMS IN BUILDINGS

Developers, industries, and other large private-sector water users are beginning to ask: How can greywater be safely managed without using municipal systems? Can rainwater be used for drinking? Are local authorities warming up to alternative approaches?

Whereas capturing rainfall was once a charming throwback idea, primarily of interest to environmentalists, today it is serious business with safer modern holding tanks and advanced purification systems. Strict regulations are evolving around how we dispose of water after use. Back in the day, farmers and others thought of wastewater as a private matter for a landowner, but in modern times, local governments consider actions that have impact beyond your property or family. If water is managed incorrectly, hazardous substances drain into aquifers or local waterbodies. As densities increase or operations expand, all water systems are subject to inspection.

In growing cities, infrastructure is often too old, crumbling, and inadequate for the demands placed on it. Many North American municipalities have peculiar trunk designs that did not anticipate modern populations. When stormwater rises quickly, greater volumes of untreated sewage is dumped into waterbodies, increasing the danger of *Escherichia coli* (*E. coli*) bacterial infection, leading to closure of popular beaches in the summer.

For this reason, expanding cities prefer downpipes from rooftops to be disconnected from storm sewers and require developers to restrict storm and sewage flows. Meanwhile, the one-hundred-year-old system is being gradually upgraded, water and sewage rates are increasing, and rainwater harvesting technologies are returning to popularity.

SOLAR PANELS CAPTURE RAINWATER FOR HIGHRISE

The Edith Green-Wendell Wyatt Federal Building is a half-million-square-foot, eighteen-story office tower located in downtown Portland, Oregon. It was renovated about five years ago, and mechanical, electrical, data, fire, and life safety systems were upgraded. In addition, a deep energy retrofit was undertaken that included building envelope improvements and solar panels on a thirteen-thousand-square-foot roof.

The panels were angled to optimize their yield from the sun, and mechanical designers realized this might be ideal for capturing rainfall. Their hunch penciled out in dollars, so they installed a 165,000-gallon rain catchment tank in an unused underground rifle range beneath the building. The complex needs about ten thousand gallons per day in the winter. In summer, cooling towers and irrigation increase this to between twelve thousand and seventeen thousand gallons. The new system created cost savings and reduced the building's environmental impact. Total water collection is around 626,500 gallons each year, 85 percent of the site's requirements.[6]

LEARNING CENTERS IN ST. LOUIS AND AMHERST, MASSACHUSETTS

The Tyson Learning Center, a field station for Washington University in St. Louis, and the Hitchcock Center, a field-trip facility in Amherst, Massachusetts, for younger students learning about water cycles, are both water neutral. They collect enough rainwater for all their requirements and use rain gardens for outgoing infiltration to groundwater.[7]

Dan H., the architect of the St. Louis project, says they submitted a design and the local government said, "You can't do that." They organized a meeting with the public works director and walked him through the technology and received plan approval. Later, building inspectors ordered a stoppage. They were also unfamiliar with the systems, and another meeting was called with the technology again explained patiently before work could continue.

In Texas, an elementary school is saving about five hundred thousand gallons each year by collecting rainfall from its forty-five thousand square feet of rooftops. Water is collected in two fifteen-thousand-gallon cisterns for bathrooms and in a twelve-thousand-gallon cistern for watering the school's landscaping.[8] Mark G., a director for the company that supplies the equipment, says that rainwater for drinking is not yet permitted in most parts of the United States, but that the technology to do it safely is completely proven.

Research companies offer different predictions about how quickly rainwater-harvesting technology use will grow in the coming decade; however, most see increases of at least 5 percent each year, with several predicting an industry that exceeds one billion US dollars in sales annually.

WATER TREATMENT PLANTS

Chapter 12 showed that heating and cooling energy is now being extracted from sewage pipes. Another high-impact nexus between water and energy involves cutting emissions and saving energy during water-treatment operations. Dr. Luxmy Begum is an engineering expert in this field. "Nature's mesmerizing beauty has always held a special place in my heart," says Dr. Begum. "I graduated in civil engineering from the highest-ranked engineering institute in India (IIT, Madras). But my heart was not in civil engineering. I decided to pursue my master's degree in environmental engineering instead. I attended the Asian Institute of Technology (AIT) in Thailand, where I was introduced to the wonderful world of environmental preservation and protection. For the first time, it felt like home. I discovered that protecting the environment was worth spending my career on and the very purpose of my life."

Her passion for environmental preservation has catalyzed a career filled with project achievements and, as she does in chapter 18, with sharing her cutting-edge knowledge on how medium-to-large wastewater treatment plants can move toward energy efficiency and energy neutrality.

Chapter Eighteen

Toward Energy Neutrality

Energy-Saving Strategies and Emerging Technologies for Wastewater Treatment Plants[1]

Luxmy Begum, PhD, PEng, PMP

A wastewater-treatment and resource-recovery specialist, Dr. **Luxmy Begum**, PEng, holds a PhD in Environmental Engineering from the University of Tokyo and a master's from AIT in Thailand. She has more than twenty years of professional experience in the water- and waste-management sector. This includes undertaking treatment-plant upgrades, technology selection, evaluation, and procurement for water, wastewater, and anaerobic digestion plants.

At one time, adopting energy-saving strategies at wastewater treatment plants (WWTPs) was rarely considered; but today, it has become crucial. With increasing populations, limited energy sources, ever-increasing utility costs, and looming climate change threats, there is no way to ignore it.

Moreover, recent advancements in resource recovery and reuse technologies have caused a shift in how we think about sewage and sludge. Rather than viewing them solely as waste to be disposed of, we are now seeing them as potential sources from which valuable resources can be recovered. With this new outlook, the goal is to have treatment plants that are not only energy-efficient but also energy-neutral and even energy-positive, if feasible. The objective of this chapter is to review and summarize various energy-saving strategies and evolving technologies that can help WWTPs achieve these objectives.[2]

ENERGY AND WASTEWATER

Energy represents a substantial cost in wastewater treatment, as it is required at almost all stages in the treatment process. With pumps, motors, and other

169

equipment operating twenty-four hours a day, seven days a week, water and wastewater facilities can be among the largest consumers of energy in a community. But if we look at the energy content of the wastewater, the findings can be surprising. As noted by the Water Environment Federation (in 2016), wastewater is a renewable resource embedded with an abundance of energy, including thermal at about 80 percent, hydraulic at about 19 percent, and chemical at about 1 percent.

Typical wastewater contains nearly five times the amount of energy needed to treat it. The wastewater industry could harness that energy and eliminate not only its net consumption but also generate excess energy for other users at a competitive price. Wastewater facilities have the potential to produce the energy needed not only to treat water but also to help heat and power the cities that depend on them.

ENERGY-SAVING STRATEGIES FOR WASTEWATER TREATMENT PLANTS

According to data derived from the Water Environment Energy Conservation Task Force (Wisconsin Utilities & Focus on Energy, 2020) the most energy-consuming processes for wastewater treatment plants are the activated sludge aeration system, which accounts for approximately 54 percent of the energy used, and the pumping process, approximately 14 percent. Just by improving these two processes, plants can achieve significant energy savings.

SAVING ENERGY BY OPTIMIZING THE AERATION SYSTEM

The aeration process can account for the largest energy demand of any operation at a facility. But focusing on improving energy consumption in the following areas can result in total aeration energy savings of 30 to 70 percent (SAIC, 2006):

- diffuser type and configuration
- blower type and configuration
- dissolved oxygen monitoring and control technologies

The wastewater aeration system (blower and diffuser assembly) is the number one energy consumer in a plant. Simply running the blower with variable frequency drives (VFDs), which offer more speeds and increase optimization options, compared with conventional one-speed blowers, can result in energy savings of 15 to 50 percent (SAIC, 2006). An energy-efficient blower can also make a significant difference. Turbo blowers offer much higher efficiency (70 to 80 percent efficient) compared with conventional blowers (45 to 65 percent).

Another effective strategy is to automate the monitoring of dissolved oxygen and use control technology that can maintain the dissolved oxygen level of the aeration tank(s) at a preset point by varying the rate of the airflow to the aeration system.

The development of fine-bubble diffuser technology offers significant reductions in aeration energy consumption compared with mechanical and coarse-bubble aeration. Fine-bubble technologies have applications for all sizes of wastewater treatment facilities. The proportion of energy savings will be similar regardless of facility size, and this technology has gained a high level of acceptance within the industry. It has been estimated that using fine-bubble diffusers can reduce aeration energy from 25 percent to as much as 75 percent (SAIC, 2006).

In addition, changes in the number of diffusers and their configuration can further increase energy efficiency. A common approach is to use tapered aeration to reduce the rate of the oxygen supplied along the length of a basin (EPA, 2010). In this approach, more diffusers are placed at the inlet to the basin, where the organic loading is highest, and fewer diffusers are placed along the basin's length. Tapered aeration better matches the actual oxygen demand across the basin: more air is moved to the head of the basin where it is most needed, with less air sent to the end of the basin where the food-to-microorganisms ratio is lower, thereby saving energy.

Other strategies for optimizing the aeration system include:

- operating fewer aeration tanks
- idling an aeration tank during low-flow periods
- reducing the airflow to the aeration tanks during low-load periods (usually nights and weekends)
- recycling backwash water during off-peak power demand periods
- using multiple smaller tanks to step the system into operation, rather than having only two large tanks; this approach allows for energy-efficient operation from start-up to design-flow conditions.

SAVING ENERGY BY OPTIMIZING
PUMP SYSTEM EFFICIENCY

Pumps are applied at many points in a wastewater treatment plant. To improve pump efficiency, proper sizing is crucial. Rather than sizing the pump for peak flow, which occurs infrequently, facilities should consider replacing large-capacity pumps (that operate intermittently) with smaller-capacity pumps that will operate for longer periods and closer to their best efficiency point (i.e., the flow point at which the pump operates at its highest or optimum efficiency). Other strategies to improve the efficiency of the pump system are:

- lowering the pumping capacity to better match system demands;
- installing a parallel system for highly variable flows;
- selecting more efficient motors; and
- installing better flow control.

VFDs are the most flexible and efficient way to control flow. Since equipping all pumps with VFDs can be costly, one strategy is to install a VFD for the duty pump, as it runs most of the time.

ADOPTING EMERGING TECHNOLOGIES AND INNOVATIVE PATHWAYS

The new goal for WWTPs is to reduce energy consumption while maximizing energy recovery and production. WWTPs can achieve this by adopting the following:

Enhanced Primary Treatments Processes[3]—This allows more organics to be sent to anaerobic digestion (or other energy-recovery process) and means that less aeration will be required in the secondary treatment because of early-stage organic removal. The following are some of the benefits of enhanced primary treatment technologies:

- Improved primary treatment reduces the organic load on subsequent biological processes; less biological process capacity is required in the downstream.
- Power consumption is reduced because the reduced organic load downstream means substantially less aeration is required.
- Greenhouse gas (GHG) emissions are reduced because the amount of CO_2 released into the air due to aeration falls considerably.
- Collected solids can be sent further for enhanced energy and nutrient recovery.
- Organic material can be diverted during peak energy demand periods and sent back in the early-morning hours.

- Plant capacity is increased cost-effectively, as more of the total suspended solids (TSS) and biochemical oxygen demand (BOD) is removed before secondary treatment effectively increases the capacity of the WWTP without having to make expensive capital upgrades to the aeration tanks.

LESS ENERGY-INTENSIVE
BIOLOGICAL TREATMENT PROCESSES

Greater focus has been placed on developing technologies that can cut down the cost of aeration during wastewater treatment and nutrient recovery. Some promising emerging technologies have already been developed and tested in full-scale, full-plant demonstrations.[4] This approach reduces aeration demand and increases energy recovery. Phototrophic technologies[5] can increase the chemical energy of a wastewater through CO_2 fixation during growth and carbon storage. This excess energy can then be recovered later through various energy-recovery processes, such as anaerobic digestion.

ENERGY RECOVERY VIA
SLUDGE TREATMENT TECHNOLOGIES

In the past, wastewater sludge and biosolids were considered as wastes to be disposed of. However, with recent focus on circular economy and GHG emission reduction, we are compelled to look for different alternatives for biosolids disposal/treatment. We discovered that reusing biosolids is one of the most sustainable ways to achieve a circular economy that eliminates waste, recovers renewable energy, and enhances the environment at the same time. Moreover, the recent emerging breakthrough technologies have also opened doors for different energy and resource recovery potential for biosolids. Some of these existing and emerging energy recovery technologies from sewage sludge and biosolids are anaerobic digestion, pyrolysis, gasification, hydrothermal liquefaction (HTL), and other supercritical gasification/oxidation.

Anaerobic digestion—Biogas can be generated via anaerobic digestion of sewage sludge and can be collected and used in boilers at facilities, converted to heat and electricity using combined heat and power engines, or cleaned and upgraded to renewable natural gas (RNG) for injection into the natural gas distribution grid.

Pyrolysis—The pyrolysis process involves thermochemical decomposition of biosolids at elevated temperatures (500 to 800°C). The pyrolysis process takes place in the absence of oxygen, and end products are process gas (py-gas/pyrogas), oil, and biochar. Pygas can be further converted to energy, and biochar can be used for soil amendment (EDI, 2017).

Gasification—Gasification is a thermochemical process that converts bio-solids at a high temperature (500 to 1,500°C) into ash and "syngas." Syngas can be further converted into either energy, fuel, or synthetic biofuels. The ash generated as a by-product can be used as a soil amendment.

Hydrothermal liquefaction (HTL)—Hydrothermal liquefaction (HTL) is a thermochemical process that converts many types of biomass, including sludge and biosolids, into an energy-dense bio-crude oil and co-products like gases, aqueous phase, and char.

Conclusion—The adoption of a combination of various energy-saving strategies plus multiple energy-saving and energy-producing process units such as enhanced primary treatment, co-digestion with high-strength waste, advanced high-rate anaerobic digestion, and other energy-producing sludge treatment technologies, side-stream treatments like deammonification, and additional alternative-energy sources in the plant, such as solar panels or wind turbines, may be required for WWTPs to make significant progress toward achieving energy neutrality or energy-positive status.

Chapter Nineteen

Urban Planning, Government Priorities, and Oxygen

BF Nagy

Our challenge for the next ten years can be stated simplistically as ending fossil fuel use and water pollution caused by transportation, buildings, industry, biodiversity mismanagement, and poor agriculture practices. Obviously, governments at all levels have a critical role to play in these areas.

Urban planners and municipal governments should make it a priority to redesign city transportation and buildings, advance and enforce new building codes and green building standards, and phase out exclusionary zoning. Let's make it easy to build laneway housing, create distributed solar and microgrids, and use public lands for geothermal. We must demand greater decarbonization and water protection ambition from developers during the site plan process and prioritize bike lanes, trails, public spaces, public transit investment, public charging, food waste planning, recycling, circular economy systems, and partner with developers to create wastewater energy recovery systems.

Urban planners can also collaborate with other government levels, institutions, and private entities to create seamless grid integration for microgrids, modern district energy, vehicle-to-grid projects, virtual power plants, and other energy demand response programs and adopt a climate-resilience lens for planning.

States and provinces need to rein in ridiculous disinformation, homeowner opposition to solar, dealer opposition to electric vehicles (EVs); improve grid power market mechanisms; incentivize renewables, solar, wind, and batteries; and drive fossil fuels out of the energy system. States should adopt modern EV rules, develop charging networks, support electrified fleets and buses, phase out the sale and use of combustion vehicles, adopt new building standards, invest in new training for inspectors, work with the federal government on industrial policy that creates battery, semiconductor, and EV manufacturing jobs, and ban fossil fuels in all buildings.

Federal governments must immediately end subsidies and pointless "research" by fossil fuel companies, create carbon taxes, get tougher on regulating their profits and disinformation, and consider nationalizing them. Let's pass powerful laws to halt bottom trawling, overfishing, and trophy hunting. We must protect animals, marine life, oceans, waterbodies, forests, and biodiversity (set aside 30 percent, at a minimum).

Planners at all government levels, please declare a climate emergency, create policy and regulatory frameworks with teeth, budget for investment in effective enforcement, and change political donation laws. National governments need to pass specific legislation to expressly protect the rights of peaceful climate protesters who are not causing damage or egregious disruption and provide free speech training to police authorities everywhere. Fossil fuel companies must be minimized in the policy-making process and barred from COP and other major climate-decision conferences. Federal governments should provide leadership and work with other government levels and utilities to beef up incentives for well-sealed buildings and heat pumps, distributed solar, and EVs. All governments should get their own houses in order, accelerating electrification of government and institutional buildings and fleets and water system protection.

The climate crisis imperative means that mainstream media journalists and social media influencers are critical players in the battle. We must tell better, more colorful stories about the success of our clean energy heroes and present accurate information about climate change causes and proven solutions.

Aviation and shipping is regulated by international bodies, which have been laissez-faire and negligent. National governments should create new rules wherever they can, sending strong signals to these bodies and the distance-shipping/travel industry to accelerate decarbonization timelines. France has made it illegal to fly short distances. Airlines and shipping companies, take note.

TOP FIVE LIST

My last book provided Top Ten lists of priority sustainability actions for people in particular situations or occupations. These can now be replaced with the simplified Top Five list that follows, which can be used by all of us in the critical 2024–2035 period.

1. VOTE	☑	Vote in all elections at every level and only for people who support rapid climate action.
2. RIDE	🚗	Electrify your personal and work rides, switching to transit, bike, eBike, electric car, electric truck etc.
3. BUILDING	🏠	At home and work, swop out fossil fuels with electric heat pumps, tight envelopes, water conservation, solar panels.
4. LEARN	📰	Learn about green living, myths, incentives.
5. SHARE	🖥	Share what you learn in a helpful way, not shaming.

Figure 19.1. Top 5 Climate Actions
Figure created by Climate Solution Group.

RECKLESS COURAGE AND OXYGEN

As a young newspaper delivery boy in the city, I think I neglected my customers because I was more interested in reading the articles. And while camping with my family in the outdoors at a place called Killbear Provincial Park in Ontario, I finally began writing at the age of thirteen.

If you're familiar with paintings by the Group of Seven, then you have an idea of what the Killbear area looks like: a mix of pine and maple forests, wild blueberries, rounded cliffs of salmon-colored feldspar, glittering quartz, and stately grey granite, hidden coves, secret beaches, and magical forest nooks. The place was teeming with wildflowers, squirrels, foxes, rabbits, and turtles.

I was at the beach on my own one morning, trying to make blades of grass squeak out musical notes and get stones to skip off the waves, the way my brothers could. After a while, I clambered up a nearby rocky hill, passing other kids collecting blueberries, catching praying mantises, or sitting on beach towels, counting sailboats in the harbor. From their pink feldspar perches, they could take in two or more bays and beaches, loads of the little islands planted with small colored cottages, docks, rafts, and canoes.

I heard some splashing and shouting nearby and went to investigate. On that side of the hill was a cliff face that dropped steeply to the swimming area below. The water was clear, and you could see that it was deep. Some older boys were throwing themselves off different precipices, plunging into the cool bay, while their younger siblings squealed and cheered. I was enthralled. It seemed like they were flying about one hundred feet before they hit the water, but in reality, it was about twenty-five feet. I watched them for a while and dreamed of a day when I would be brave like them.

Then it happened. A spindly little girl, about ten years of age, walked by me over to the cliff's edge, dropped her towel on the rocks, and jumped over the side, like there was nothing to it. She was so small, she seemed to fly even longer. My mouth was hanging open, and this changed everything. If she could do it—it was like your brother arriving for Thanksgiving dinner in an electric car, or your neighbor putting solar panels on her roof.

I asked one of the older boys if I could jump too. He said, "Sure, why not? Just make sure you push off with your feet. You don't want to hit any rocks on the way down." So, I did it. The first time, I jumped from one of the lower ledges. The third time, from the very top. It seemed unbelievable that I was actually doing this. It was also unlikely, according to my brothers when I told them, and they made me prove it. By the end of the day, we'd spent hours there. I was like a tour guide, showing them where to make their first attempts before they advanced to my superior level. They saw me differently that day.

The next morning, the new me rose at dawn and climbed out of the tent, blinking like an alpaca and mindlessly watching the ash-colored clouds brighten minute by minute. I took my journal and tiptoed, barefoot, down the gravel camp road, then padded along a cool dirt path to the beach. I scrambled up the same cliff just in time for the sun to peak above the trees and play across the lazy morning waves. Nobody was around. Birds were tweeting, frogs were croaking, and I wrote a poem. I convinced myself I would be the next William Shakespeare.

Could it have happened in a city full of concrete? Sure, but in this case, creative inspiration was probably enhanced by a combination of reckless confidence and being under the influence of clean, highly oxygenated air. We visited Killbear a few more times when I was young, and as an adult, I took my own family there. Decades after I wrote that first poem, I returned there, all by myself.

I'd been working nonstop since graduation, messing up relationships and career phases, and I was still chasing deadlines for magazine stories. The pandemic had begun, and there was talk of a recession. I decided to slow down and sold my beloved Kew Beach condominium in Toronto. I did it quickly without any immediate thought to the future. Not even a rental apartment. It was summertime, and I was deliciously homeless and rudderless, and I felt more free than I had in years. I loitered at the family cottage. When they needed my bedroom, I took a tent, my laptop, and an external six-hour battery to Killbear. I was seeking an inspiring, unencumbered place, where I could think.

I did go back to the rocky hill that birthed my passion, but the best moment for me on that trip happened in a different part of the park. I chose one of the sections near an inlet where the forest is dense all the way to the shore. I wandered through the trees during the day and discovered, under a big oak tree, a ten-foot stretch of pebble and sand that faced west across the water. The nights had been warm, so I returned to that spot just before sunset with my lawn chair, computer, and a sandwich.

I sunk the lawn chair into the sand of the small beach so that when I sat down my feet would be in the warm water as it lapped gently onto the shore. I nibbled on my cheese and tomato while the blush-colored sun spread itself delicately across the ripples reaching toward the horizon. My marriage had ended some years earlier, and my son was now grown. I was feeling lonely and sentimental.

I'm not someone who gives up. In baseball or tennis, it's never over until it's over. You focus on the next few moments and don't dwell on the last strikeout or missed shot. I knew that, in a few days, I would return to civilization and busy myself with finding a new town, a new home, new friends, a new tennis club, and a whole new set of projects and plans. But for now, I was not a father, uncle, brother, author, filmmaker, journalist, or webinar host. I was nothing but a guy in a lawn chair, in a provincial park, eating a sandwich.

However, rather than being nothing in a dingy apartment that drains the life out of me, I was nothing on a beautiful sandy shore, with the lake water babying my feet, swaddled by a sweet-smelling forest, with a soft breeze ruffling my hair, and a bounteous crimson sunset warming my heart. I was nothing and alone and yet somehow fulfilled. In that moment, I reaffirmed my purpose and my re-entry into the world of the living, with help from the power of nature. I rededicated myself to the battle for safe water, clean air, and healthy food, secure in the knowledge that they are worth the fight it will take to save them.

Humanity doesn't give up either. Our families, our friends, our communities, a huge majority of the hopeful millions who are still alive today, along with the real experts who brought you the passages in this book, are all counting on the rest of us to help in small ways to speed up the transition to a clean home planet. Because rather than draining us, the natural beauty on Earth refills our reservoirs, makes us kinder and more thoughtful,

> "Now I see the secret of the making of the best persons, It is to grow in the open air and to eat and sleep with the earth."
> —Walt Whitman, from "Song of the Open Road," 1856

inspires our creativity, recharges our batteries, and restores our courage. And there is nothing, absolutely nothing, that is more valuable than that, anywhere in the universe.

Notes

CHAPTER 2. THE REALITIES OF THE CLIMATE IMPERATIVE

1. Bell, James, Jacob Poushter, Moira Fagan, and Christine Huang, "In Response to Climate Change, Citizens in Advanced Economies Are Willing To Alter How They Live and Work," Pew Research, https://www.pewresearch.org/global/2021/09/14/in-response-to-climate-change-citizens-in-advanced-economies-are-willing-to-alter-how-they-live-and-work/.

2. NASA, "Do scientists agree on climate change?" https://climate.nasa.gov/faq/17/do-scientists-agree-on-climate-change/.

3. Porter, Michael E., Forest L. Reinhardt, et al., "Risk Management—Climate Business | Business Climate—From the Magazine (October 2007)," *Harvard Business Review*, https://hbr.org/2007/10/climate-business-_-business-climate.

4. "Big business sees the promise of clean energy," *The Economist*, 2017, https://tinyurl.com/2b6tpxn7.

5. Ramkumar, Amrith, "Clean-Energy Funding Stayed Strong in Weak Market: Investors bet on Washington to juice spending on green industries in 2023," *Wall Street Journal*, December 30, 2022, https://www.wsj.com/articles/clean-energy-funding-stayed-strong-in-weak-market-11672352953.

6. SWNS News delivery platform by Google, August 22, 2023, "Most Gen Z and Millennials base purchases on a brand's mission: poll," https://nypost.com/2023/08/22/most-gen-z-millennials-base-purchases-on-brands-mission-poll/.

Parmelee, Michele, "Making waves: How Gen Zs and millennials are prioritizing—and driving—change in the workplace," 2023 Deloitte Insights, https://www2.deloitte.com/uk/en/insights/topics/talent/recruiting-gen-z-and-millennials.html.

7. Metcalf, Tom, and Pei Yi Mak, "These Billionaires Made Their Fortunes by Trying to Stop Climate Change: They're among the first to profit from climate solutions—but they won't be the last," Bloomberg.com, January 22, 2020, https://www.bloomberg.com/features/2020-green-billionaires/#xj4y7vzkg.

8. 60/215 = 28 percent maintenance savings, 53/214 = 25 percent utility savings, 50/168 = 30 percent cleaning savings: Energystar.gov, "Quantifiable Costs Savings—Compared to typical buildings, high-performing buildings save: $0.60 per square foot on operations and maintenance expenses annually, $0.50 per square foot on janitorial expenses annually, $0.53 per square foot on utility expenses annually," https://www.energystar.gov/buildings/save_energy_commercial_buildings/comprehensive_energy_management/business_case.

Constellation.com, "Commercial Real Estate: Building Operating Cost Breakdown . . . $2.15 per square foot on repairs and maintenance, $2.14 on utilities, $1.68 on cleaning," https://tinyurl.com/3r9s4r2p.

9. See chapter 3 of this book.

CHAPTER 3. 100% RENEWABLES AND THE SIX-YEAR PAYBACK

1. Completed in 2018, https://web.stanford.edu/group/efmh/jacobson/Articles/I/TownsCities.pdf.

2. Ibid.

3. Completed in 2015, with a handful more later. Extensive press coverage in 2017. https://thesolutionsproject.org/what-we-do/inspiring-action/why-clean-energy/.

4. Mark Z. Jacobson, "100% Clean, Renewable Energy and Storage for Everything," 2020, https://web.stanford.edu/group/efmh/jacobson/WWSBook/WWSBook.html; Mark Z. Jacobson, "No Miracles Needed," 2023, https://web.stanford.edu/group/efmh/jacobson/WWSNoMN/NoMiracles.html.

5. Lazard, "Levelized cost of energy," 2021, https://www.lazard.com/research-insights/levelized-cost-of-energy-levelized-cost-of-storage-and-levelized-cost-of-hydrogen-2021/.

6. Mark Z. Jacobson, "No Miracle Tech Needed: How to Switch to Renewables Now and Lower Costs Doing It," 2022, https://thehill.com/opinion/energy-environment/3539703-no-miracle-tech-needed-how-to-switch-to-renewables-now-and-lower-costs-doing-it/.

7. Ibid.

8. Mark Z. Jacobson, "No, we don't need 'miracle technologies' to slash emissions—we already have 95 percent," 2021, https://thehill.com/opinion/energy-environment/554605-no-we-dont-need-miracle-technologies-to-slash-emissions-we-already/.

9. Mark Z. Jacobson, "No Miracle Tech Needed."

10. Ibid.

11. Mark Z. Jacobson, "No, we don't need 'miracle technologies.'"

12. KRWG | by Sierra Club, "PNM cites ETA in proposing 100% solar and storage to replace nuclear," 2021, https://www.krwg.org/local-viewpoints/2021-04-16/pnm-cites-eta-in-proposing-100-solar-and-storage-to-replace-nuclear.

13. Mark Z. Jacobson and Mark A. Delucchi, "A Plan for a Sustainable Future," *Scientific American*, 2009, https://web.stanford.edu/group/efmh/jacobson/Articles/I/sad1109Jaco5p.indd.pdf.

14. http://web.stanford.edu/group/efmh/jacobson/Articles/I/CountryGraphs/CO2 ChangesWithWWS.pdf.

15. https://web.stanford.edu/group/efmh/jacobson/Articles/I/NuclearVsWWS.pdf.

16. https://web.stanford.edu/group/efmh/jacobson/Articles/I/WWSDiagram.pdf.

17. https://web.stanford.edu/group/efmh/jacobson/Articles/I/NonEnergySolutions.pdf.

18. Ibid.

19. https://web.stanford.edu/group/efmh/jacobson/Articles/Others/19-CCS-DAC .pdf.

20. Robert W. Howarth and Mark Z. Jacobson, "How Green Is Blue Hydrogen?" *Energy Science & Engineering* 9, Issue10 (October 2021): 1676–87, https://doi .org/10.1002/ese3.956; https://onlinelibrary.wiley.com/doi/full/10.1002/ese3.956.

21. Mark Z. Jacobson, "Effects of biomass burning on climate, accounting for heat and moisture fluxes, black and brown carbon, and cloud absorption effects," *JGR Atmospheres* 119, Issue 14 (July 27, 2014): 8980–9002, https://doi.org/10.1002 /2014JD021861

CHAPTER 4. SELLING OUR HOME TO THE HIGHEST BIDDER

1. Dr. Michael E. Mann, *The New Climate War* (New York: Public Affairs Books), https://michaelmann.net/. Dr. Mann won the Nobel Peace Prize in 2007 along with numerous other authors of the Intergovernmental Panel on Climate Change.

2. Joseph Winters, "The EPA has a controversial new plan to clean up power plants," Canary Media, May 12, 2023, https://www.canarymedia.com/articles/utilities /is-carbon-capture-viable-in-a-new-rule-the-epa is asking-power-plants-to-prove-it/.

3. MIT, "cancelled and inactive projects," http://sequestration.mit.edu/tools/proj ects/index_cancelled.html.

4. Ron Bousso, "Big Oil doubles profits in blockbuster 2022," Reuters, February 2023, https://www.reuters.com/business/energy/big-oil-doubles-profits-block buster-2022-2023-02-08/.

5. National Research Defense Council (NRDC), "Lies the Koch Brothers Tell," August 2023, https://www.nrdc.org/stories/lies-koch-brothers-tell; Jane Mayer, "Kochland Examines the Koch Brothers' Early, Crucial Role in Climate-Change Denial," *The New Yorker*, August 2019, https://www.newyorker.com/news/daily -comment/kochland-examines-how-the-koch-brothers-made-their-fortune-and-the -influence-it-bought; Phoebe Cooke, "Fossil Fuel Groups 'Spent Millions' on Social Media Ads Spreading Climate Disinformation During COP27," Desmog.com, January 2023, https://www.desmog.com/2023/01/19/fossil-fuel-groups-spent-millions -on-social-media-ads-spreading-climate-disinformation-during-cop27/.

6. Simon Black, Ian Parry, and Nate Vernon, "Fossil Fuel Subsidies Surged to Record $7 Trillion," International Monetary Fund, August 2023, https://www.imf .org/en/Blogs/Articles/2023/08/24/fossil-fuel-subsidies-surged-to-record-7-trillion, https://www.imf.org/en/Publications/WP/Issues/2023/08/22/IMF-Fossil-Fuel-Subsi dies-Data-2023-Update-537281.

CHAPTER 5. FALSE SOLUTIONS FROM THE GAS INDUSTRY

1. BBC, "Big oil vs. the world, part 3: Delay," 2023, https://vimeo.com/734313866 (password BIGOIL2022). Last accessed August 9, 2023.

2. International Energy Agency, "World gas production, 1973–2020," 2021, https://www.iea.org/data-and-statistics/charts/world-natural-gas-production-by-region-1973-2020. Last accessed August 9, 2023.

3. Energy Information Administration, "U.S. gas marketed production," U.S. Department of Energy, 2023, https://www.eia.gov/dnav/ng/hist/n9050us2A.htm. Last accessed August 9, 2023.

4. Sharma, Vipul, "Big oil profits hit record high levels in 2022," Energy, April 26, 2023, https://www.visualcapitalist.com/cp/big-oil-profits-reached-record-high-levels-in-2022/. Last accessed August 9, 2023.

5. BBC, "Big oil vs. the world, part 3: Delay"; King, Bob, "Sierra Club faces gas-cash fallout," Politico, February 2, 2023, https://www.politico.com/story/2012/02/sierra-club-wrestles-with-gas-cash-aftermath-072581. Last accessed August 9, 2023.

6. Plumer, Brad, "Obama says 'fracking' can be a 'bridge' to a clean-energy future. It's not that simple," *Washington Post*, January 29, 2014, https://www.washingtonpost.com/news/wonk/wp/2014/01/29/obama-says-fracking-offers-a-bridge-to-a-clean-energy-future-its-not-that-simple/. Last accessed August 9, 2023.

7. Howarth, Robert W., Renee Santoro, and Anthony Ingraffea, "Methane and the greenhouse gas footprint of gas from shale formations," *Climatic Change Letters* 106 (April 21, 2011): 679–90, doi: 10.1007/s10584-011-0061-5.

8. IPCC, "Climate change 2021: The physical science basis," Intergovernmental Panel on Climate Change, 2021, https://www.ipcc.ch/report/sixth-assessment-report-working-group-i/. Last accessed August 9, 2023.

9. Zeller, Tom, Jr., "Studies Say Natural Gas Has Its Own Environmental Problems," *New York Times*, April 11, 2011, https://www.nytimes.com/2011/04/12/business/energy-environment/12gas.html.

10. CNBC, "Energy Matters," April 12, 2011, http://mediacenter.tveyes.com/downloadgateway.aspx?UserID=39625&MDID=666715&MDSeed=8619&Type=Media. Last accessed August 9, 2023.

11. Zeller, "Studies Say Natural Gas Has Its Own Environmental Problems."

12. American Gas Alliance, "Experts speak on Howarth. Nat Gas is a cleaner alternative. Get the facts here. www.anga.us/howarth." Advertisement on Google for twenty-two months in 2011 through 2013. Robert Howarth citation index (August 2023). Google Scholar, https://scholar.google.com/citations?user=sHsS7nAAAAAJ&hl=en. Last accessed August 8, 2023.

13. Howarth, Robert W., "Methane emissions from the production and use of gas," *EM Magazine* (December 2022): 11–16.

14. Gordon, Deborah, Frances Reuland, Daniel J. Jacob, John R. Worden, Drew Shindell, and Mark Dyson, "Evaluating net life-cycle greenhouse gas emissions intensities from gas and coal at varying methane leakage rates," *Environmental Research Letters* 8, No. 9 (July 2023): 10.1088/1748-9326/ace3db; Tabuchi, Hiroko, "Leaks Can Make Gas as Bad for the Climate as Coal, a Study Says," *New*

York Times, July 13, 2023, https://www.nytimes.com/2023/07/13/climate/natural-gas
-leaks-coal-climate-change.html. Last accessed August 9, 2023.

15. Gaventa, Jonathan, and Maria Pastukhova, "Gas under pressure as IEA
launches net-zero pathway," Energy Monitor, May 18, 2021, https://energymonitor
.ai/policy/net-zero-policy/gas-under-pressure-as-iea-launches-net-zero-pathway. Last
accessed August 9, 2023.

16. Figueres, Christiana, "Gas, like coal, has no future as the world wakes up to
climate emergency," *South China Morning Post*, August 29, 2021, https://www.scmp
.com/comment/opinion/article/3146479/gas-coal-has-no-future-world-wakes-climate
-emergency/. Last accessed August 9, 2023.

17. Milulka, Justin, "Major fossil fuel PR group is behind pro-hydrogen push,"
DeSmog, December 9, 2020, https://www.desmogblog.com/2020/12/09/fti-consult
ing-fossil-fuel-pr-group-behind-europe-hydrogen-lobby. Last accessed August 9, 2023.

18. Howarth, Robert W., and Mark Jacobson, "How green is blue hydrogen?"
Energy Science and Engineering 9, No. 10 (October 2021): 1676–87, doi: 10.1002
/ese3.956.

19. New York Climate Action Council, "Members," https://climate.ny.gov/Re
sources/Climate-Action-Council. Last accessed August 9, 2023.

20. Howarth, Robert W. "Methane emissions from fossil fuels: Exploring
recent changes in greenhouse-gas reporting requirements for the State of New
York," *Journal of Integrative Environmental Sciences* 17, No. 3 (2020): i–viii, doi:
10.1080/1943815X.2020.1789666.

21. New York State Climate Action Council, "Final scoping plan," https://climate
.ny.gov/resources/scoping-plan/. Last accessed August 9, 2023.

22. Romano, Matteo C., et al., "Comment on 'How green is blue hydrogen?'"
Energy Science & Engineering 10 (2022): 1944–54, https://onlinelibrary.wiley.com
/doi/pdf/10.1002/ese3.1126?trk=public_post_comment-text.

23. Howarth, Robert W., and Mark Jacobson, "Reply to comment on 'how green
is blue hydrogen?'" *Energy Science and Engineering* 10, No. 7 (July 2022): 1955–60,
doi: 10.1002/ese3.1154.

24. National Grid, "What is hydrogen?" February 23, 2023, https://www.national
grid.com/stories/energy-explained/what-is-hydrogen. Last accessed August 9, 2023.

25. Weiss, Tessa, Chathurika Gamage, Thomas Koch Blank, Genevieve Lillis,
and Alexandra Jardine Wall, "Hydrogen reality check: All 'clean hydrogen' is not
equally clean," Rocky Mountain Institute, October 4, 2022, https://rmi.org/all-clean
-hydrogen-is-not-equally-clean/. Last accessed August 9, 2023.

26. Stokes, Leah C., "Before we invest billions in this clean fuel, let's make
sure it's actually clean," *New York Times*, April 14, 2023, https://www.nytimes
.com/2023/04/14/opinion/hydrogen-fuel-tax-credit-climate-change.html. Last accessed August 9, 2023.

27. Krupnick, Alan, and Aaron Bergman, "Incentives for clean hydrogen in the Inflation Reduction Act," Resources for the Future, November 9, 2022, https://www.rff
.org/publications/reports/incentives-for-clean-hydrogen-production-in-the-inflation
-reduction-act/. Last accessed August 9, 2023.

28. ACAPMag, "Gas isn't transitional but a 'forever fuel' says Chevron executive Jeff Gustavson," ACAPMag, August 9, 2023, https://acapmag.com.au/2023/08/gas-isnt-transitional-but-a-forever-fuel-says-chevron-executive-jeff-gustavson/. Last accessed August 10, 2023.

29. Cebon, David, "Pursuing the hydrogen economy as a climate solution will be a big mistake," Inside Track, February 11, 2021, https://greenallianceblog.org.uk/2021/02/11/pursuing-the-hydrogen-economy-as-a-climate-solution-will-be-a-big-mistake/. Last accessed August 9, 2023; Lowes, Richard, Bridget Woodman, and Jamie Speirs, "Heating in Great Britain: An incumbent discourse coalition resists an electrifying solution," *Transitions* 37 (December 2020): 1–17, https://doi.org/10.1016/j.eist.2020.07.007; Grant, Alex, and Paul Martin, "Hydrogen is big oil's last grand scam," Clean Technica, https://www.jadecove.com/research/hydrogenscam/. Last accessed August 9, 2023.

30. EESI, "Fact sheet energy storage," Environmental and Energy Study Institute, February 22, 2019, https://www.eesi.org/papers/view/energy-storage-2019. Last accessed August 9, 2023.

31. Storrow, Benjamin, "Can gas be fossil free? This utility says it can," Climate Wire, Politico Pro, April 20, 2022, https://subscriber.politicopro.com/article/eenews/2022/04/20/can-gas-be-fossil-free-this-utility-says-it-can-00026444. Last accessed August 9, 2023; Kurmayer, Nikolaus J., "Heating homes with hydrogen fails on economic and climate merit: Report," Euractiv, November 18, 2021, https://www.euractiv.com/section/energy/news/heating-homes-with-hydrogen-fails-on-economic-and-climate-merit-report/. Last accessed August 9, 2023; Prieto, Olivia, and Mike Henchen, "Low-carbon fuels have a limited role to play in New York's buildings," Rocky Mountain Institute, May 25, 2022, https://rmi.org/low-carbon-fuels-have-a-limited-role-to-play-in-new-yorks-buildings/. Last accessed August 9, 2023; Collins, Leigh, "Revealed: What 18 independent studies all concluded about the use of hydrogen for heating," Recharge, June 17, 2022, https://www.rechargenews.com/energy-transition/revealed-what-18-independent-studies-all-concluded-about-the-use-of-hydrogen-for-heating/2-1-1240962. Last accessed August 9, 2023; Lowes, Richard, and Devon Cebon, "'Wrong side of history' | Wake up to the hype around green hydrogen for heating," Energy Transition, Recharge, August 24, 2022, https://www.rechargenews.com/energy-transition/wrong-side-of-history-wake-up-to-the-hype-around-green-hydrogen-for-heating/2-1-1282365. Last accessed August 10, 2023.

32. Rosenow, Jan, "Is heating homes with hydrogen all but a pipe dream? An evidence review," *Joule* 6, No. 10 (October 19, 2022): 2225–28, https://doi.org/10.1016/j.joule.2022.08.015; Baldwin, Sara, Dan Esposito, and Hadley Tallackson, "Assessing the viability of hydrogen proposals: Considerations for state utility regulators and policy makers," Energy Innovation, March 2022, https://energyinnovation.org/wp-content/uploads/2022/03/Assessing-the-Viability-of-Hydrogen-Proposals.pdf. Last accessed August 9, 2023.

33. Engineering with Rosie, "Hydrogen in the gas network, interview with Paul Martin," https://youtu.be/vrKvj2MHLVw. Last accessed August 10, 2023.

34. Pearl, Larry, "Hydrogen blends higher than 5% raise leak, embrittlement risks for gas pipelines: California PUC," Utility Dive, July 22, 2022. https://www.utility dive.com/news/hydrogen-blends-higher-than-5-percent-raise-leak-embrittlement -risks/627895/. Last accessed August 10, 2023.

35. Ocko, Ilissa, and Steven Hamburg, "New research reaffirms hydrogen's impact on the climate, provides consensus," Environmental Defense Fund, July 19, 2023, https://blogs.edf.org/energyexchange/2023/07/19/new-research-reaffirms-hydrogens -impact-on-the-climate-provides-consensus/. Last accessed August 10, 2023.

36. Los Angeles Times Editorial Board, "Editorial: Hoping fossil fuel giants will see the light on climate hasn't worked: Change only comes with mandates and force." *Los Angeles Times*, July 21, 2023, https://www.latimes.com/opinion /story/2023-07-21/editorial-its-not-enough-to-be-frenemies-with-fossil-fuel-compa nies-we-have-to-kick-them-to-the-curb/. Last accessed August 10, 2023.

CHAPTER 6. OPPORTUNITY COSTS AND DISTRACTIONS

1. Bloomberg, "A Record $495 Billion Invested in Renewable Energy in 2022," February 2023, https://about.bnef.com/blog/a-record-495-billion-invested-in-renew able-energy-in-2022/.

2. Selig, Katie, "Judge rules in favor of Montana youths in landmark climate deci- sion," *Washington Post*, August 14, 2023, https://www.washingtonpost.com/climate -environment/2023/08/14/youths-win-montana-climate-trial/.

3. Horsman, Paul, "Spills, explosions and looming disasters: The last thing the world needs is more oil," https://www.greenpeace.org/international/story/53638 /spills-explosions-no-more-oil/.

4. Nakat, Ghiwa, "Yemen: Collaboration triumphs to protect Red Sea, no thanks to Big Oil," Greenpeace.com, August 12, 2023, https://www.greenpeace.org/interna tional/story/61302/yemen-red-sea-safer-oil-risk-averted-un-make-polluters-pay/.

5. Harvey, Fiona, "Oil and gas facilities could profit from plugging methane leaks, IEA says," *Guardian*, February 23, 2022, https://www.theguardian.com/environ ment/2022/feb/23/oil-and-gas-facilities-could-profit-from-plugging-methane-leaks -iea-says.

6. Nobel, Justin, "Where Does All The Radioactive Fracking Waste Go?" Desmog International, April 21, 2022, https://www.desmog.com/2021/04/22/lotus-llc-radioac tive-fracking-waste-disposal-texas/?utm_source=DeSmog%20Weekly%20Newsletter.

7. MacMillan, Douglas, "The dangerous business of dismantling America's aging nuclear plants," *Washington Post*, May 13, 2022, https://www.washingtonpost.com /business/2022/05/13/holtec-oyster-creek-nuclear-plant-cleanup/.

8. Drysdale, Sam, "State rejects effort to dump nuclear plant waste into Cape Cod Bay," CBS News, July 24, 2023, https://www.cbsnews.com/boston/news/cape-cod -bay-nuclear-plant-waste-dump-holtec-pilgrim/.

9. These costs and the chart data are based on Lazard 2023: https://www.lazard .com/media/nltb551p/lazards-lcoeplus-april-2023.pdf.

10. Bergengruen, Vera, "Authorities Fear Extremists Are Targeting U.S. Power Grid," *Time*, January 9, 2023, https://time.com/6244977/us-power-grid-attacks-extremism/.

11. Harker, Joe, "Concrete tomb filled with deadly nuclear waste is leaking as it's starting to crack," Unilad, August 12, 2023, https://www.unilad.com/news/us-news/runit-dome-nuclear-waste-leaking-pacific-ocean-540419-20230812; Coleen, Jose, Kim Wall, and Jan Hendrik Hinzel, "This dome in the Pacific houses tons of radioactive waste—and it's leaking," *The Guardian*, July 3, 2015, https://www.theguardian.com/world/2015/jul/03/runit-dome-pacific-radioactive-waste.

12. Feldman, Nicole, "The steep costs of nuclear waste in the U.S.," Stanford, July 2018, https://earth.stanford.edu/news/steep-costs-nuclear-waste-us.

13. Ferris, Nick, "Why a new era for US nuclear looks unlikely," Energy Monitor, May 2023, https://www.energymonitor.ai/sectors/power/why-a-new-era-for-us-nuclear-looks-unlikely/.

14. Plumer, Brad, "U.S. Nuclear Comeback Stalls as Two Reactors Are Abandoned," *New York Times*, July 31, 2017, https://www.nytimes.com/2017/07/31/climate/nuclear-power-project-canceled-in-south-carolina.html.

15. Amy, Jeff, "The first US nuclear reactor built from scratch in decades enters commercial operation in Georgia," *AP News*, July 2023, https://apnews.com/article/georgia-power-nuclear-reactor-vogtle-9555e3f9169f2d58161056feaa81a425.

16. McClearn, Matthew, "Governments, utilities and the nuclear industry hope small modular reactors will power Canada's future. Can they actually build one?" *The Globe and Mail*, July 17, 2021, https://www.theglobeandmail.com/business/article-governments-utilities-and-the-nuclear-industry-hope-small-modular/.

CHAPTER 7. HERE COMES THE SUN

1. Dale, Larry, Michael Carnall, Max Wei, Gary Fitts, and Sarah Lewis McDonald, "Assessing the Impact of Wildfires on the California Electric Grid," State of California Energy Commission, August 2018, https://www.energy.ca.gov/sites/default/files/2019-11/Energy_CCCA4-CEC-2018-002_ADA.pdf.

2. Jeong, Andrew, "California shuts down major hydroelectric plant after Lake Oroville water levels fall amid severe drought, climate change," *The Washington Post*, August 6, 2021, https://www.washingtonpost.com/nation/2021/08/06/california-drought-oroville-powerplant/. Last accessed May 1, 2023.

3. Murphy, Caitlin, Trieu Mai, Yinong Sun, Paige Jadun, Matteo Muratori, Brent Nelson, and Ryan Jones, *Electrification Futures Study: Scenarios of Power System Evolution and Infrastructure Development for the United States*, January 1, 2021, https://www.nrel.gov/docs/fy21osti/72330.pdf. Last accessed July 19, 2022.

4. Larson, Eric, et al., *Net-Zero America: Potential Pathways, Infrastructure, and Impacts,* Net-Zero America, October 29, 2021, https://netzeroamerica.princeton.edu/the-report/. Last accessed July 19, 2022.

5. Barbose, Galen, Salma Elmallah, and Will Gorman, *Behind-the-Meter Solar+Storage: Market data and trends*. ETA Publications, July 2021, https://eta-publications.lbl.gov/sites/default/files/btm_solarstorage_trends_final.pdf/. Last accessed August 8, 2022.

6. Mims Frick, Natalie, Snuller Price, Lisa Schwartz, Nichole Hanus, and Ben Shapiro, *Locational Value of Distributed Energy Resources*, ETA Publications, February 2021, https://eta-publications.lbl.gov/sites/default/files/lbnl_locational_value_der_2021_02_08.pdf.

7. California Public Utilities Commission, "CPUC President's Ruling Orders Edison to Submit Information on Undergrounding Transmission Line in Chino Hills," CPUC, July 2, 2012, https://docs.cpuc.ca.gov/PUBLISHED/NEWS_RELEASE/169967.htm. Last accessed July 19, 2022; "Decision Granting the City of Chino Hills' Petition for Modification of Decision 09-12-044 and Requiring Undergrounding of Segment 8a of the Tehachapi Renewable Transmission Project," CPUC, July 11, 2013; "Tehachapi Renewable Transmission Project," Chino Hills, https://www.chinohills.org/1000/Tehachapi-Renewable-Transmission-Project/. Accessed July 19, 2022.

8. Deign, Jason, "So, What Exactly Are Virtual Power Plants?" Greentech Media, October 22, 2020, https://www.greentechmedia.com/articles/read/so-what-exactly-are-virtual-power-plants/. Last accessed July 19, 2022.

9. Chen, Olivia. "What Is Grid Edge?" Greentech Media, January 1, 2017, https://www.greentechmedia.com/articles/read/what-is-the-grid-edge#gs.__dDpTo/. Last accessed July 19, 2022.

10. Guccione, Leia, and Laurie Guevara-Stone, "Why colleges are big believers in microgrids," GreenBiz, October 17, 2013, https://www.greenbiz.com/article/why-colleges-are-big-believers-microgrids. Last accessed July 19, 2022.

11. Way, Rupert, Matthew Ives, Penny Mealy, and J. Doyne Farmer, "Empirically grounded technology forecasts and the energy transition," *Joule* 6, No. 9 (September 21, 2022): 2057–82; Nemet, Greg, "How Solar Became Cheap," 2019, https://www.howsolargotcheap.com/. Last accessed May 1, 2023.

12. CaliforniaDGStats, "Statistics and Charts," June 30, 2022, https://www.californiadgstats.ca.gov/charts/. Last accessed August 8, 2022.

13. Nyberg, Michael, "2021 Total System Electric Generation," California Energy Commission, https://www.energy.ca.gov/data-reports/energy-almanac/california-electricity-data/2020-total-system-electric-generation/. Last accessed August 8, 2022.

14. "Australia installs record-breaking number of rooftop solar panels," CSIRO, Press Release, May 13, 2021, https://www.csiro.au/en/news/all/news/2021/may/australia-installs-record-breaking-number-of-rooftop-solar-panels/. Last accessed August 8, 2022.

15. Executive Summary, "International Country Analysis—AUS," International–U.S. Energy Information Administration (EIA), March 18, 2022, https://www.eia.gov/international/analysis/country/AUS/. Last accessed August 8, 2022.

16. "What are DERMS and what are the benefits of using them?" Wood Mackenzie, September 9, 2020, https://www.woodmac.com/news/editorial/derms-for -growth/. Last accessed August 9, 2022.

17. "Solar Power Purchase Agreements | SEIA," Solar Energy Industries Association, https://www.seia.org/research-resources/solar-power-purchase-agreements. Last accessed August 9, 2022.

18. "Self-Generation Incentive Program (SGIP)," California Public Utilities Commission, https://www.cpuc.ca.gov/industries-and-topics/electrical-energy/demand -side-management/self-generation-incentive-program. Last accessed August 9, 2022.

19. "FERC Order No. 2222: Fact Sheet," FERC, September 17, 2020, https:// www.ferc.gov/media/ferc-order-no-2222-fact-sheet.

20. Kennedy, Ryan, "U.S. grid-scale and residential energy storage hit installation records in Q3," *pv magazine*, December 15, 2022, https://pv-magazine-usa .com/2022/12/15/u-s-grid-scale-and-residential-energy-storage-hit-installation-rec ords-in-q3/. Last accessed May 1, 2023.

21. "Energy Storage Legislation," California Public Utilities Commission, https:// www.cpuc.ca.gov/industries-and-topics/electrical-energy/energy-storage. Last accessed August 10, 2022.

22. "How California Is Driving the Energy Storage Market Through State Legislation," Climate Group, April 2017, https://www.theclimategroup.org/sites/default /files/2020-11/under2_coalition_case_study_etp_california.pdf. Last accessed August 10, 2022.

23. "New Data Shows Growth in California's Clean Electricity Portfolio and Battery Storage Capacity," California Energy Commission, May 25, 2023, https://www .energy.ca.gov/news/2023-05/new-data-shows-growth-californias-clean-electricity -portfolio-and-battery. Last accessed June 1, 2023.

24. "Self Generation Incentive Program Evaluation Reports," California Public Utilities Commission, https://www.cpuc.ca.gov/industries-and-topics/electrical-en ergy/demand-side-management/self-generation-incentive-program/self-generation -incentive-program-evaluation-reports. Last accessed August 10, 2022.

25. "Repowering Clean Gigawatt-Scale Potential for Residential Solar & Battery Storage in Los Angeles," Sunrun, https://www.sunrun.com/sites/default/files /repowering-clean-sunrun.pdf. Last accessed August 13, 2022.

26. "What are DERMS?"

27. Purdom, Sophie, and Kim Zou, "Lessons from Plaid for a future energy unicorn," Climate Tech VC | Substack, March 11, 2022, https://climatetechvc.substack .com/p/-lessons-from-plaid-for-a-future. Last accessed August 13, 2022.

28. Penn, Ivan, "Its Electric Grid Under Strain, California Turns to Batteries," *New York Times*, September 3, 2020, https://www.nytimes.com/2020/09/03/business /energy-environment/california-electricity-blackout-battery.html. Last accessed September 26, 2022.

29. Misbrener, Kelsey, "Residential solar batteries helped California's grid with peak power during heatwave," Solar Power World, September 9, 2022, https://www .solarpowerworldonline.com/2022/09/customer-sited-solar-batteries-340-mw-peak -power-one-day-california/. Last accessed September 26, 2022.

30. Rewiring America, "Your guide to the Inflation Reduction Act—Rewiring America," Rewiring America, 2022, https://www.rewiringamerica.org/IRAguide. Last accessed October 22, 2022.

31. Shibu, Sherin, and Shana Lebowitz, "Microsoft Performance Reviews: How They Work Under Satya Nadella," *Business Insider*, November 23, 2020, https://www.businessinsider.com/microsoft-performance-reviews-how-they-work-under-satya-nadella-2019-8. Last accessed August 13, 2022.

CHAPTER 8. CHANGE

1. Wheat, Steve, "To love the water, itself," Halfwaydownthestairs.net, September 2022, https://halfwaydownthestairs.net/2022/09/01/to-love-the-water-itself-by-steve-wheat/#anchor.

2. Counterpointresearch.com, "Global EV Sales Up 32% YoY in Q1 2023 Driven by Price War," June 2023, https://www.counterpointresearch.com/global-ev-sales-q1-2023.

3. Counterpointresearch.com, "Global electric vehicle market highlights," August 2023, https://www.counterpointresearch.com/global-electric-vehicle-market-share/.

4. Baranzini, Andrea, Stefano Carattini, and Martin Péclat, "What drives social contagion in the adoption of solar photovoltaic technology?" London School of Economics and Political Science, July 2017, https://www.lse.ac.uk/granthaminstitute/publication/what-drives-social-contagion-in-the-adoption-of-solar-photovoltaic-technology/.

5. "Barriers to Electric Vehicle Adoption in 2022," Exro.com, August 2023, https://www.exro.com/industry-insights/barriers-to-electric-vehicle-adoption-in-2022.

6. "Building the 2030 National Charging Network," NREL Study, June 2023, https://www.nrel.gov/news/program/2023/building-the-2030-national-charging-network.html.

7. Lewis, Michelle, "Here's how many EV chargers the US has—and how many it needs," Electrek.com, January 2023, https://electrek.co/2023/01/09/heres-how-many-ev-chargers-the-us-has-and-how-many-it-needs/.

8. Nagy, BF, "The Mobility Revolution," *Ontario Technologist*, March 2023, 24–27, https://www.mydigitalpublication.com/publication/?m=30652&i=787698&p=24&ver=html5.

CHAPTER 9. POLICY AND PLANNING FOR ELECTRIC VEHICLE GRID INTEGRATION

1. Alliance for Automotive Innovation, "Guiding Principles," retrieved from: https://www.autosinnovate.org/about/advocacy/EV%20Infrastructure%20Initiative.pdf.

2. New York State Energy Research and Development Authority (NYSERDA), "Benefit-Cost Analysis of Electric Vehicle Deployment in New York State," NYSERDA Report Number 19-07, 2019. Prepared by E3, ICF, and MJ Bradley. nyserda.ny.gov/publications; M. J. Bradley & Associates (now ERM) has performed cost-

benefit analysis for EV adoption in nineteen states, including Arizona, Connecticut, Florida, Michigan, Minnesota, New Mexico, Nevada, and Pennsylvania. Their analyses consistently find that both EV drivers and utility ratepayers benefit from EV adoption, especially when rates and programs encourage load shifting.

3. Hawaiian Electric et al., "Electrification of Transportation Strategic Roadmap," March 2018, 34–35. Filed in Hawaii Public Utilities Commission Docket No. 2016-0168, March 29, 2018. https://www.hawaiianelectric.com/documents/clean_en ergy_hawaii/electrification_of_transportation/201803_eot_roadmap.pdf; M. J. Bradley & Associates, "Electric Vehicle Cost-Benefit Analysis: Connecticut," 20. https://mjbradley.com/sites/default/files/CO_PEV_CB_Analysis_FINAL_13apr17.pdf.

4. Environmental Defense Fund and ERM, "Electric Vehicle Market Update," September 2022. https://www.edf.org/media/worldwide-electric-vehicle-investments -will-grow-more-626-billion-2030-new-report.

5. Bloomberg NEF, "Race to net zero: Pressures of battery boom in five charts," July 15, 2022. https://www.bloomberg.com/professional/blog/race-to-net-zero-pres sures-of-the-battery-boom-in-five-charts/.

6. Burton, Mark, "Race for Cheaper EVs May Face Years-Long Setback on Metal Costs," Bloomberg US Edition, May 31, 2022. https://www.bloomberg.com /news/articles/2022-06-01/battery-price-increase-could-keep-electric-cars-expensive -for-years?leadSource=uverify%20wall.

7. See, for example, ERM (MJ Bradley & Associates), "Plug-in Electric Vehicle Cost-Benefit Analysis: Nevada," "Plug-in Electric Vehicle Cost-Benefit Analysis: Minnesota," and "Plug-in Electric Vehicle Cost-Benefit Analysis: Florida."

8. Whited, Melissa, et al., "Electric vehicles are driving electric rates down," June 2023. https://www.synapse-energy.com/ev-rate-impacts.

9. Wolfram, Paul, Stephanie Weber, Kenneth Gillingham, and Edgar G. Herwich, "Pricing indirect emissions accelerates low-carbon transition of US light vehicle sector," *Nature Communications* 12, 7121 (2021). https://doi.org/10.1038/s41467 -021-27247-y.

10. McLaren, Joyce, et al., *Emissions Associated with Electric Vehicle Charging: Impact of Electricity Generation Mix, Charging Infrastructure Availability, and Vehicle Type,* National Renewable Energy Laboratory Technical Report NREL/ TP-6A20-64852, prepared under Task No. VTP2.0100, April 2016. www.nrel.gov/ publications; Knobloch, Florian, et al. "Net emission reductions from electric cars and heat pumps in 59 world regions over time," *Nature Sustainability* 3 (March 23, 2020): 437–47. doi:10.1038/s41893-020-0488-7.

11. Lutsey, Nic, and Dale Hall, "Effects of battery manufacturing on electric vehicle life-cycle greenhouse gas emissions," The International Council on Clean Transportation, February 2020. https://theicct.org/wp-content/uploads/2021/06 /EV-life-cycle-GHG_ICCT-Briefing_09022018_vF.pdf; Hausfather, Zeke, "Factcheck: How electric vehicles help tackle climate change," CarbonBrief, May 13, 2019. https://www.carbonbrief.org/factcheck-how-electric-vehicles-help-to-tackle -climate-change/.

12. See Ryan, Nancy, et al., "Best Practices for Sustainable Commercial EV Rates and PURPA 111(d) Implementation," National Association of Regulatory Util-

ity Commissioners, 2022. https://pubs.naruc.org/pub/55C47758-1866-DAAC-99FB-FFA9E6574C2B.

13. Blair, Brittany, Garrett Fitzgerald, and Carolyn Dougherty, *The State of Managed Charging in 2021*, Smart Electric Power Alliance, November 2021, p. 8.

14. Smart Electric Power Alliance, "Residential Electric Vehicle Rates that Work," November 2019; Cook, Jonathan, Candice Churchwell, and Stephen George, *Final Evaluation of San Diego Gas & Electric's Plug-in Electric Vehicle Pricing and Technology Study*, Nexant, Inc., February 20, 2014. https://www.sdge.com/sites/default/files/SDGE%20EV%20%20Pricing%20%26%20Tech%20Study.pdf.

15. Compliance Filing of Southern California Edison Company (U 338-E), San Diego Gas & Electric Company (U 902 E), and Pacific Gas and Electric Company (U 93 E) Pursuant to Ordering Paragraph 2 of Decision 16-06-011, filed with the California Public Utilities Commission Docket No. R18-12-006, April 1, 2022.

16. Cappers, Peter, et al., *Snapshot of EV-Specific Rate Designs Among U.S. Investor-Owned Electric Utilities*, Lawrence Berkeley National Laboratory, April 2023.

17. For examples, see Energy + Environmental Economics, *Rate Designs Harnessing Vehicle Grid Integration Technology: Novel Rate Designs for Aggregator Enabled Smart Charging*, May 2021.

18. Blair, *The State of Managed Charging in 2021*; Opinion Dynamics, "PG&E Electric Vehicle Automated Demand Response Study Report," CALMAC Study ID PGE0469.01, February 2022. See appendix A for a survey of managed charging programs across the United States.

19. Smart Electric Power Alliance, "Residential Electric Vehicle Rates that Work," 18–20 and 22–23; Ibid., 25 ff.

20. Opinion Dynamics, "PG&E Electric Vehicle," 69–74.

21. State of New York Public Service Commission, "Order Approving Managed Charging Programs with Modifications," Case 18-E-0138, July 14, 2022.

22. Opinion Dynamics, "PG&E Electric Vehicle," 54–56.

23. Smart Electric Power Alliance, "Residential Electric Vehicle Rates that Work," 16 and 28–29.

24. Lipman, Timothy, Alissa Harrington, and Adam Langton, "Total Charge Management of Electric Vehicles," California Energy Commission, December 2021. https://www.energy.ca.gov/sites/default/files/2021-12/CEC-500-2021-055.pdf.

25. "National Grid Launches Off-Peak Electric Vehicle Charging Program for Massachusetts Customers," National Grid, July 7, 2022. https://www.nationalgridus.com/News/National-Grid-Launches-Off-Peak-Electric-Vehicle-Charging-Program-for-Massachusetts-Customers/.

26. "Electric Vehicle Charging Incentives," conEdison. https://www.coned.com/en/save-money/rebates-incentives-tax-credits/rebates-incentives-tax-credits-for-residential-customers/electric-vehicle-rewards.

27. DTE Press Release, "BMW joins Ford, GM in DTE Energy's Smart Charge Program, integrating electric vehicle charging with their net zero carbon emissions goals," May 9, 2022. https://ir.dteenergy.com/news/press-release-details/2022/BMW-joins-Ford-GM-in-DTE-Energys-Smart-Charge-Program-integrating-electric-vehicle-charging-with-their-net-zero-carbon-emissions-goals/default.aspx.

28. This approach is currently being piloted in Connecticut, Maryland, and Colorado. In North Carolina, Duke Energy has initiated a one-year pilot expressly aimed at demonstrating that submetering via vehicle telematics can eliminate the need for a dedicated second meter or a networked smart charger.

29. Utility-managed charging programs should maximize the pool of potential participants by allowing participation via both telematics and networked L2 chargers, according to Blair, "The State of Managed Charging in 2021," 44–45; and Opinion Dynamics, "PG&E Electric Vehicle," 61.

30. For additional recommendations, see Vehicle Grid Integration Council, "V2X Bidirectional Charging Systems: Best Practices for Service Connection or Interconnection," August 2022. https://static1.squarespace.com/static/5dcde7af8ed96b403d8aeb70/t/62fd4c3cfc19490ee68d71eb/1660767294489/VGIC-Special-Initiative-2022.pdf.

CHAPTER 10. THE ENERGY STORAGE CHALLENGE

1. Harlow, Jessie E., et al., "A Wide Range of Testing Results on an Excellent Lithium-Ion Cell Chemistry to be used as Benchmarks for New Battery Technologies," *Journal of The Electrochemical Society* 166, No, 13 (2019); Update 2022: Aiken, C. P., et al., "Li[Ni0.5Mn0.3Co0.2]O2 as a Superior Alternative to LiFePO4 for Long-Lived Low Voltage Lithium-Ion Cells," Published on behalf of The Electrochemical Society by IOP Publishing Limited, *Journal of The Electrochemical Society* 169, No. 5 (May 9, 2022). Soc. 169 050512DOI 10.1149/1945-7111/ac67b5, PDF: https://iopscience.iop.org/article/10.1149/1945-7111/ac67b5/pdf, https://iopscience.iop.org/article/10.1149/1945-7111/ac67b5 (June 2023).

2. Ibid.

3. Eldesoky, A., et al., "Impact of Graphite Materials on the Lifetime of NMC811/Graphite Pouch Cells: Part II. Long-Term Cycling, Stack Pressure Growth, Isothermal Microcalorimetry, and Lifetime Projection," Published on behalf of The Electrochemical Society by IOP Publishing Limited, *Journal of The Electrochemical Society* 169, No. 1 (January 5, 2022). Soc. 169 010501DOI 10.1149/1945-7111/ac42f1, https://iopscience.iop.org/article/10.1149/1945-7111/ac42f1/meta (June 2023).

The team concluded that NMC811/graphite cells will benefit from an enormous lifetime boost when operated at a limited UCV of 4.06 V, where decades-long lifetimes can be achieved at 20°C to 30°C, if the best graphites are selected.

4. More lab and research information: Some researchers might be interested in this additional information on the Dalhousie lab and research into ways to ensure that lithium-ion batteries can be efficient, capable, safe, and inexpensive. Data charts, photos, graphics, and references for coulombic efficiency of small parasitic reactions between the charged electrode materials and the electrolyte, battery microcalorimetry, lithium-ion differential thermal analysis, x-ray photoelectron spectroscopy, and pouch bag studies to understand electrodes/electrolyte reactions and electrolyte additives can all be found on our research site at https://www.dal.ca/diff/dahn/research/adv_diagnostics/hpc_additive_studies.html.

5. According to a 2020 report in "The Driven" (https://thedriven.io/2020/06/05/battery-day-why-teslas-single-crystal-cathode-is-important/), Tesla CEO and co-

founder Elon Musk has said that the EV maker's best-selling Model 3 is designed to last one million miles, but current battery chemistry used by Tesla only lasts for up to five hundred thousand miles (800,000 km)—or, fifteen hundred cycles. "Model 3 drive unit & body is designed like a commercial truck for a million mile life. Current battery modules should last 300k to 500k miles (1500 cycles). Replacing modules (not pack) will only cost $5k to $7k," wrote Musk on Twitter in 2019.

NOVONIX of Australia has filed patents for single-crystal, very-long-distance cathodes for lithium-ion batteries. Chris Burns (Tesla) and David Stephens (Dahn Lab) founded NOVONIX in 2013, based on their work with Professor Dahn. The 2023 announcement of the patent filing included a suggestion from Burns that its DPMG single-crystal technology can be created at reduced cost and with better performance.

CHAPTER 11. AVIATION ELECTRIFICATION

1. A version of this article appeared in Illuminem Voices, May 3, 2022. https://illuminem.com/illuminemvoices/aviation-is-a-major-climate-problem-but-electrification-and-biofuels-are-solutions.

CHAPTER 12. REINVENTING BUILDINGS AND COMMUNITIES

1. Bertram, Nick, Steffen Fuchs, Jan Mischke, Robert Palter, Gernot Strube, and Jonathan Woetzel, "Modular Construction: From Projects to Products," McKinsey & Company, June 2019, https://www.mckinsey.com/capabilities/operations/our-insights/modular-construction-from-projects-to-products.

2. *The Economist*, "The rise of 3D-printed houses: Your next home could be a printout," August 2021, https://www.economist.com/science-and-technology/the-rise-of-3d-printed-houses/21803667.

3. Ideas and content in this chapter were informed over the years by site visits and story assignments from mainstream media and also from architecture and engineering publications. Two of these should be specifically acknowledged: *Plumbing Engineer* magazine—American Society of Plumbing Engineers, and the *Ontario Technologist* magazine, Toronto. Thanks also to the Clean Air Partnership, sponsor of my urban planning sustainability webinar series under the auspices of the Federation of Canadian Municipalities.

4. Nagy, BF, "How today's green building heroes are scaling up to save our planet," *Corporate Knights* magazine, February 2023, https://www.corporateknights.com/category-buildings/how-todays-green-building-heroes-are-scaling-up-to-save-our-planet/.

See also bfnagy.com for fifteen-plus years of selected magazine features and videos on green buildings, communities, transportation, government policy, and best practices.

5. NBC TV quoting the National Oceanic and Atmospheric Administration (NOAA), https://www.nbcmiami.com/news/local/ian-is-costliest-hurricane-in-florida-history-caused-112b-in-damage-in-us-noaa/3007223.

6. Nagy, BF, "Survival Tech for Extreme Weather," PHCP pros (Engineering), December 2022, https://www.phcppros.com/articles/16572-survival-tech-for-extreme-weather.

7. From BF Nagy interview with Syd Kitson and Amy W. after the Florida hurricane for this story on the largest green building communities in North America: Nagy, "How today's green building heroes are scaling up to save our planet."

8. Video on Whisper Valley in Austin Texas, engineer Chad B., https://www.you tube.com/watch?v=-yxzVvNgBNQ.

9. "The Guide to Federal, State, and Utility Incentives for Geothermal Heat Pumps," https://dandelionenergy.com/geothermal-state-federal-tax-incentives.

10. Nagy, BF, "The Heat Beneath Our Feet," PHCP Pros, July 2023, https://www.phcppros.com/articles/17757-the-heat-beneath-our-feet.

11. City of Richmond, "Lulu Island Energy Company," September 2022, https://www.richmond.ca/city-hall/news/city2022/districtenergyexpansion2022sept26.htm.

12. Diverso Energy, "All the Benefits, None of the Risk—Geothermal Simplified," https://www.diversoenergy.com/our-projects.

13. Geosource Energy, "The Power of Geoexchange," https://www.geosource energy.com/.

14. Heat pumps, AHRI: https://www.ahrinet.org/analytics/statistics/historical -data/central-air-conditioners-and-air-source-heat-pumps.

Furnaces. AHRI: https://www.ahrinet.org/analytics/statistics/historical-data/fur naces-historical-data.

15. International Energy Agency: Yannick Monschauer, Chiara Delmastro, and Rafael Martinez-Gordon, " Global heat pump sales continue double-digit growth," March 2023, https://www.iea.org/commentaries/global-heat-pump-sales-continue -double-digit-growth. Worldwide installations of heat pumps increased by 11 percent in 2022 after a similar increase in 2021. Air-to-water models, which are compatible with typical radiators and underfloor heating systems, jumped by almost 50 percent in Europe. The agency said that heat pumps supply heat for more than one hundred million households around the world.

16. Nagy, BF, "Can We Truly Electrify America?" ASPE Engineers, August 2020, https://www.phcppros.com/articles/11967-can-we-truly-electrify-america.

17. Chung, Emily, "How Sweden electrified its home heating—and what Canada could learn," CBC, April 2023, https://www.cbc.ca/news/science/sweden-heat -pumps-1.6806799.

18. International Passive House Association, "Passive House Certification Criteria," https://passivehouse-international.org/index.php?page_id=150.

19. Wood Mackenzie, "US Distributed Energy Resource market to almost double by 2027," June 2023, https://www.woodmac.com/press-releases/us-distributed-en ergy-resource-market-to-almost-double-by-2027/.

20. Solar Energy Industries Association, "Solar Market Insight Report: 2022 Year in Review," https://www.seia.org/research-resources/solar-market-insight-report -2022-year-review.

21. Nagy, "How today's green building heroes are scaling up to save our planet."

22. https://bfnagy.com/.

23. Nagy, "How today's green building heroes are scaling up to save our planet."

24. Nagy, "How today's green building heroes are scaling up to save our planet."

25. Nagy, BF, "Is Off-Grid Living Making a Comeback?" *Plumbing Engineer* magazine, July 2020, https://www.phcppros.com/articles/11675-is-off-grid-living-making-a-comeback.

26. Energiesprong.org, "Energiesprong Countries," https://energiesprong.org.

27. *Technical description:* "Infiltration rates of the renovated buildings are extremely low, measured as low as 0.34 cubic dm/s/sq. m. at 10 Pa (0.15 CFM/SF @ 50 Pa), and typically in the range of 0.4–0.6 cubic dm/s/sq. m. at 10 Pa (0.18–0.26 CFM/SF @ 50 Pa). These infiltration rates are not quite as tight as Passive House (approximately 0.11 CFM/SF @ 50 Pa), but they are tighter than the new 2016 New York State code requirement, which is approximately 0.56 CFM/SF @ 50 Pa. To deliver sufficient ventilation, a balanced heat recovery ventilation system is installed. One design uses room-by-room ventilation instead of a central heat recovery ventilation system."

28. PHFA data provided by Tim M. of Onion Flats, Philadelphia.

29. Department of Energy, "Heat Pump Systems," https://www.energy.gov/energysaver/heat-pump-systems.

30. Nagy, BF, "Technology Disruption," *Plumbing & HVAC* magazine, November 2020, https://plumbingandhvac.ca/technology-disruption/.

31. *Chatelaine*, https://chatelaine.com/food/chefs-induction-stoves/.

32. Induction stoves: https://www.canada.ca/en/health-canada/services/publications/healthy-living/factsheet-cooking-and-indoor-air-quality.html.

33. Rocky Mountain Institute, https://rmi.org/insight/gas-stoves-pollution-health.

34. Asthma: https://www.ncbi.nlm.nih.gov/pmc/articles/PMC9819315.

35. EPA: https://www.epa.gov/indoor-air-quality-iaq/household-energy-and-clean-air.

36. World Health Organization: https://www.who.int/news-room/fact-sheets/detail/household-air-pollution-and-health.

37. US Environmental Protection Agency (EPA), "How We Use Water," https://www.epa.gov/watersense/how-we-use-water 300 gallons per day × 120 million families × 365 days = 13,140,000,000,000.

CHAPTER 13. HOCKEY, NATURAL REFRIGERANTS, AND GREENING INDUSTRY

1. Nagy, B. F., "America Goes Green," *Plumbing Engineer* magazine, American Society of Plumbing Engineers, https://www.phcppros.com/articles/16209-america-goes-green.

2. B. F. Nagy webinar presentation for city planners 2021, screen 12, https://council.cleanairpartnership.org/wp-content/uploads/2021/12/Bruce-Pres-CAP-Energy-Recovery-Dec-2021-Final.pdf.

3. Refrigerant leakage rates: https://support.accuvio.com/support/solutions/articles/4000040366-annual-leakage-rate-for-the-refrigeration-air-con-hvac-

Also https://freor.com/green-wave-refrigeration-equipment/.

4. Montreal Protocol, https://bit.ly/3jGCI1V or https://www.unep.org/ozon action/who-we-are/about-montreal-protocol.

5. In 2036, the final phase-out year, emissions will be cut by 171 MMTCO2e.

6. Environmental Protection Agency: https://bit.ly/2ZDFZrV, or https://www.epa .gov/system/files/documents/2021-09/hfc-allocation-rule-nprm-fact-sheet-finalrule.pdf.

7. AHR Expo: https://bit.ly/3EoYGyf.

8. Shecco: https://bit.ly/3mlXNAf.

9. Low-charge ammonia: https://issuu.com/shecco/docs/a21_report_final.

10. Taylor, Chuck, and Todd Allsup, "Why CO_2 is a Viable Refrigerant Alternative," Manufacturing.net, May 15, 2015, https://www.manufacturing.net/home /article/13183437/why-co2-is-a-viable-refrigerant-alternative.

11. Natural refrigerant investment payback: https://freor.com/green-wave-refrig eration-equipment/.

12. Air as refrigerant: https://www.youtube.com/watch?v=X2zkNjDL4K0.

CHAPTER 14. POLAR BIODIVERSITY AND CLIMATE

1. Carbon *capture* from the atmosphere (or dissolved in the ocean) occurs in nature through photosynthesis of plants, algae, and some bacteria, which are collectively known as primary producers. This can lead to carbon *storage* in the bodies of organisms, such as tree trunks or animals. Some is lost through respiration or rotting on death, but some may be long-term buried in sediment when it becomes carbon *sequestration*.

CHAPTER 16. CIRCULAR FOOD SYSTEMS

1. Project Drawdown, "Sector Summary: Food, Agriculture, and Land Use," https://drawdown.org/sectors/food-agriculture-land-use.

2. Ranganathan, Janet, et al. "How to Sustainably Feed 10 Billion People by 2050, in 21 Charts," World Resources Institute, December 5, 2018.

3. The big five crops include wheat, rice, maize, potatoes, and soy.

4. Kurnik, Julia, and Katherine Devine, "Innovation in Reducing Methane Emissions from the Food Sector: Side of rice, hold the methane," WorldWildLife.org, https://www.worldwildlife.org/blogs/sustainability-works/posts/innovation-in-reduc ing-methane-emissions-from-the-food-sector-side-of-rice-hold-the-methane.

5. WBCSD & OP2B, "Staple crop diversification: Why and how to diversify from the big five crops (wheat, rice, maize, potato & soy)," https://www.wbcsd.org /download/file/12605.

6. Akbarpur, Bassi, "The global rice crisis," *The Economist*, March 28, 2023, https://www.economist.com/asia/2023/03/28/the-global-rice-crisis.

7. Jensen, Paul D., and Caroline Orfila, "Mapping the production-consumption gap of an urban food system: An empirical case study of food security and resilience," *Springer Nature* 13, No. 3 (February 8, 2021): 551–70.

8. Ibid., 552.

9. "8.6% of people in the EU unable to afford proper meal," Eurostat, https://ec.europa.eu/eurostat/web/products-eurostat-news/-/ddn-20220225-1.

10. "Food insecurity a real concern among the urban poor in Sub-Saharan Africa following pandemic—new report shows," World Food Programme, March 11, 2022, https://www.wfp.org/news/food-insecurity-real-concern-among-urban-poor-sub-saharan-africa-following-pandemic-new-report.

11. Barrera, Emiliano Lopez, and Thomas Hertel, "Global food waste across the income spectrum: Implications for food prices, production and resource use," *Food Policy* 98 (January 2021), https://doi.org/10.1016/j.foodpol.2020.101874.

12. Hans, Petra, "5 ways to transform our food system to benefit people and planet," World Economic Forum, March 5, 2021, https://www.weforum.org/agenda/2021/03/5-ways-transform-food-system-sustainable/.

13. Gustavsson, Jenny, Christel Cederberg, and Ulf Sonesson, "Global Food Losses and Food Waste," Food and Agriculture Organization of the United Nations, https://www.fao.org/sustainable-food-value-chains/library/details/en/c/266053/.

14. Project Drawdown, Reduced Food Waste, https://drawdown.org/solutions/reduced-food-waste.

15. Jensen and Orfila, "Mapping the production-consumption gap," 552.

16. Clinton, Nicholas, et al. "A Global Geospatial Ecosystem Services Estimate of Urban Agriculture," *Earth's Future* 6, No. 1 (January 10, 2018): 40–60.

17. "Global carbon dioxide emissions from fossil fuels and industry were 37.12 billion metric tons ($GtCO_2$) in 2021," Statistica, November 2022, https://www.statista.com/statistics/276629/global-co2-emissions/.

18. Garcia, Eduardo, "Where's the Waste? A 'Circular' Food Economy Could Combat Climate Change," *New York Times*, September 21, 2019, https://www.nytimes.com/2019/09/21/climate/circular-food-economy-sustainable.html.

19. Shoup, Mary Ellen, "EverGrain starts commercial production of upcycled barley protein: 'The timing of us reaching scale couldn't be better,'" Food Navigator USA, July 7, 2022.

20. https://www.gdrc.org/sustdev/concepts/04-e-effi.html.

21. Garcia, "Where's the Waste?"

22. Ibid.

23. "These 16 urban farmers turned to the roofs to feed their cities," https://sustainableurbandelta.com/16-urban-rooftop-farms/.

24. Ibid.

25. "Growing Fruit and Vegetables Above a Brussels Supermarket," Karuna News, https://www.karunanews.org/story/7930/growing-fruit-and-vegetables-above-a-brussels-supermarket.

26. Wang, Lucy, "Incredible rooftop farm takes over Israel's oldest mall to grow thousands of organic vegetables," InHabitat, January 15, 2017, https://inhabitat.com/incredible-rooftop-farm-takes-over-israels-oldest-mall-to-grow-thousands-of-organic-vegetables/.

27. Ibid.

28. Singhal, Aaryaman Arjun, and Brian Lipinski, "How Food Waste Costs Our Cities Millions," World Resources Institute, April 16, 2015.

29. "Food Waste in America in 2023: Statistics + Facts," RTS, https://www.rts.com/resources/guides/food-waste-america/.

30. Ibid.

31. Foley, Jonathan, "Feeding the World," *National Geographic Magazine: The Future of Food*, https://www.nationalgeographic.com/foodfeatures/feeding-9-billion/.

32. McCauley, D. J., "Urban Agriculture Combats Food Insecurity, Builds Community," *EOS*, January 25, 2021.

33. Ibid.

34. "ICLEI—Local Governments for Sustainability, 2021," *City Practitioners Handbook: Circular Food Systems*, Bonn, Germany.

35. Ibid.

36. Ibid, 49.

37. "NYC Health + Hospitals Now Serving Culturally-Diverse Plant-Based Meals As Primary Dinner Option for Inpatients at All of Its 11 Public Hospitals," NYC Health + Hospitals, January 9, 2023, https://www.nychealthandhospitals.org/press release/nyc-health-hospitals-now-serving-plant-based-meals-as-primary-dinner-op tion-for-inpatients-at-all-of-its-11-public-hospitals/.

CHAPTER 17. WATER

1. Nagy, BF, "Water-positive buildings are bubbling up," *Plumbing Engineer* magazine, August 2018, https://www.phcppros.com/articles/8076-water-positive -buildings-are-bubbling-up.

2. US Government Accountability Office, "Freshwater: Supply Concerns Continue, and Uncertainties Complicate Planning," May 2014, https://www.gao.gov /products/gao-14-430, August 2023.

3. EWG is a Washington, DC, nonprofit with scientists on the team and mostly foundation funding, https://www.ewg.org/interactive-maps/algal_blooms/map/, August 2023.

4. EPA, https://www.epa.gov/nutrientpollution/issue, August 2023.

5. EPA, https://www.epa.gov/nutrientpollution/sources-and-solutions, August 2023.

6. Nagy, BF, "Better Water Systems for a Better World," *PHCP Pros* magazine, June 2021, https://www.phcppros.com/articles/13438-better-water-systems-for-a-bet ter-world.

7. Nagy, "Water-positive buildings are bubbling up"; Nagy, "Better Water Systems."

8. Nagy, "Better Water Systems."

CHAPTER 18. TOWARD ENERGY NEUTRALITY

1. This article was originally published in *Influents* magazine (Water Environment Association of Ontario), Fall 2022, Volume 18.

2. Other references in this chapter:

EDI Environmental Dynamics Inc., *Beneficial Reuse of Biosolids Jurisdictional Review*, 2017, Victoria, British Columbia, Canada.

Science Applications International Corporation, *Water and wastewater energy best practice guidebook*, Wisconsin, 2006.

US Environmental Protection Agency, "Evaluation of energy conservation measures for wastewater treatment facilities," EPA 832-R-10-005, Washington, DC: US Environmental Protection Agency, September 2010.

US Environmental Protection Agency, "Emerging technologies for wastewater treatment and in-plant wet weather management," EPA 832-R-12-011, 2013. Prepared for EPA by Tetra Tech Inc.

Water Environment Federation, "Sheboygan Regional WWTF's waste to energy: The good, the bad, and the ugly," Water Environment Federation Residuals and Biosolids Conference, Milwaukee, Wisconsin, April 2016.

Wisconsin Utilities and Focus on Energy, *Water & Wastewater Industry Energy Best Practices Guidebook*, Wisconsin, 2020.

Winkler, M. K. H., "Segregation of biomass in aerobic granular sludge," doctoral dissertation, 2012, chapters 1 and 4.

3. These include chemically enhanced primary treatment, ballasted flocculation/settling (e.g., Actiflo and DensaDeg processes), and fine-screen technologies like micro-screening to remove excess organics from wastewater.

4. These include membrane-aerated biofilm reactors (MABR), aerobic granular sludge processes (e.g., AquaNereda), alternative nitrogen removal methods (nitritation-detritation process [e.g., SHARON]; nitritation-deammonification process [e.g., ANAMMOX, DEMON, ANITA Mox, DeAmmon, etc.]). Also, Anaerobic Technologies (adopted as a main or side stream process), including anaerobic membrane bioreactors, anaerobic sequencing batch reactors, microbial fuel cells, and anaerobic baffled reactors. And finally, Phototrophic Technologies (adopted as a main or side stream process).

5. Prominent algae-based phototrophic wastewater treatment technologies include high-rate algal ponds, algal photobioreactors, stirred tank reactors, waste stabilization ponds, and algal turf scrubbers.

Selected Bibliography

Balchin, John. *Quantum Leaps: 100 Scientists Who Changed the World*. London: Arcturus Publishing, 2010.

Begum, Luxmy. *Water Pollution: Causes, Treatments and Solutions*. Toronto: CreateSpace, 2015.

Bloomberg, Michael, and Carl Pope. *Climate of Hope*. New York: St. Martin's Press, 2017.

de Bono, Edward. *The Greatest Thinkers*. New York: GP Putnam's Sons, 1976.

Dawkins, Richard. *The Selfish Gene*. Oxford: Oxford University Press, 1989.

Foot, David K. *Boom, Bust and Echo: How to Profit from the Coming Demographic Shift*. Toronto: Macfarlane Walter & Ross, 1996.

Gaertner, Kate. *Planting a Seed*. Vancouver: Page Two Books, 2021.

Gladwell, Malcolm. *The Tipping Point*. New York: Little, Brown and Company, 2000.

———. *Outliers*. New York: Little, Brown and Company, 2008.

Gore, Al. *An Inconvenient Truth*. New York: Rodale, 2006.

———. *An Inconvenient Sequel: Truth to Power*. New York: Rodale, 2017.

Hamilton, Tyler. *Mad Like Tesla: Underdog Inventors and Their Relentless Pursuit of Clean Energy*. Toronto: ECW Press, 2011.

Hawken, Paul. *Drawdown: The Most Comprehensive Plan Ever Proposed to Reverse Global Warming*. New York: Penguin Books, 2017.

Jacobs, Jane. *The Death and Life of Great American Cities*. New York: Random House, 1961.

Jacobson, Mark. *100% Clean, Renewable Energy and Storage for Everything*. Cambridge: Cambridge University Press, 2021.

Klein, Naomi. *This Changes Everything: Capitalism vs. the Climate*. Toronto: Alfred A. Knopf, 2014.

Mann, Michael E. *The New Climate War: The Fight to Take Back Our Planet*. New York: Hachette Book Group, 2021.

McKibben, Bill. *Wandering Home: A Long Walk Across America's Most Hopeful Landscape*. New York: Macmillan, 2014.

Nagy, BF. *The Clean Energy Age: A Guide to Beating Climate Change*. Lanham, MD: Rowman & Littlefield, 2018.

Naisbitt, John. *Megatrends: Ten New Directions Transforming Our Lives*. New York: Warner Books, 1984.

Pratt, Fletcher. *All About Famous Inventors and Their Inventions*. New York: Random House, 1955.

Sussman, Amanda. *The Art of the Possible: A Handbook for Political Activism*. Toronto: McClelland & Stewart, 2007.

Suzuki, David, and Holly Dressel. *Good News for a Change: How Everyday People are Helping the Planet*. Vancouver: Greystone Books, 2003.

Toffler, Alvin. *Future Shock*. Toronto: Bantam Books, 1971.

Index

About the Contributors

WILLIAM ERNEST (BILL) MCKIBBEN

StoryWorkz.

Described as "America's most famous environmentalist," Bill McKibben is an author and journalist who has written more than a dozen books and hundreds of feature magazine articles about the environment. He has organized several movements, including campaigns which motivated top-level financial people to make high impact decisions affecting energy investment, and also 350 .org's grassroots efforts, which resulted in fifty-two hundred simultaneous demonstrations in 181 countries in 2009, and at least seven thousand events in 188 countries in 2010. In the 2020s, he is mobilizing people worldwide through the Third Act campaign. His books include *The End of Nature*, published in 1989, and *The Flag, the Cross, and the Station Wagon*, published in 2022, and he serves as the Schumann Distinguished Professor in Residence at Middlebury College in Vermont. In 2014, he was awarded the Right Livelihood Prize, sometimes called the "alternative Nobel," in the Swedish Parliament. He's also won the Gandhi Peace Award and received honorary degrees from nineteen colleges and universities. He has been inducted into the literature section of the American Academy of Arts and Sciences and was named one of the one hundred most important global thinkers by *Foreign Policy* magazine in 2009. He was educated at Harvard College in 1978 and served as editor of *The Harvard Crimson* newspaper. After graduation, he worked for five years for *The New Yorker* magazine before resigning in protest and, in 1987, moving with his family to a remote place in the southeastern Adirondacks of upstate New York, and later to Vermont. He remains a frequent contributor to *The New York Times, The Atlantic, Harper's, Orion, Mother Jones, The American Prospect, The New Yorker, The New York Review of Books, Granta, National Geographic, Rolling Stone, Adbusters,* and *Outside.* He is also a board member and contributor to *Grist.*

PROFESSOR MARK Z. JACOBSON, STANFORD UNIVERSITY

Professor Mark Z. Jacobson and his Stanford research group have advised the Bush, Obama, and Biden administrations, the US Congress, several state governments, and beyond. They have provided renewables roadmaps for fifty-three towns and cities, seventy-four metropolitan areas, all fifty US states, and 139 countries. Professor Jacobson has addressed the top international climate conferences, founded important educational groups, created high-impact campaigns with celebrities, and appeared on *The Late Show with David Letterman.* At UCLA, Jacobson noted that carbon dioxide was a leading cause of global warming and rapid melting of Arctic sea ice. He developed the first fully coupled online model that accounted for all major feedbacks among major atmospheric processes. In 2001, he calculated the number of wind turbines needed for the United States to satisfy the Kyoto Protocol. In 2008, his review paper (work from Graeme Hoste) made waves, concluding that 100 percent wind-water-solar (WWS) by 2030 was technically and economically possible. That year, he and Berkeley's Dr. Mark A. Delucchi published a widely noted piece in *Scientific American* on 100 percent renewables. In 2011, he co-founded the non-profit Solutions Project with Marco Krapels, Mark Ruffalo, and Josh Fox. Leonardo DiCaprio and a group of other celebrities joined with them to promote renewables to the public, governments, and top corporations. In 2015, he was the lead author of two peer-reviewed papers examining the feasibility of transitioning each of the fifty states to a 100 percent WWS energy system and ensuring grid reliability in the process. These provided support for US House Resolution 540 (2015), New York Senate Bill S5527, his presentation at COP21 in Paris, and, later, the Green New Deal, supported by more than three hundred elected US lawmakers.

PROFESSOR ROBERT HOWARTH, CORNELL UNIVERSITY

Cornell's Professor Robert Howarth is an internationally renowned research scientist, professor of ecology and environmental biology, journal editor, and advisor to governments. *Time* magazine named him in its "50 People Who Matter" 2011 Person of the Year issue. He was a participant at COP21, COP 24, and COP26, and, in the early 2020s, worked on New York State's Climate Action Council. He has published over 220 scientific papers, reports, and book chapters, and his work has been cited more than eighty-two thousand times in the peer-reviewed literature, making him one of the most-cited environmental scientists in the world. He and his life and research partner Roxanne Marino were instrumental in the acceptance of biogeochemistry as an important field of research, and Howarth was the founding editor of the journal *Biogeochemistry*, serving as Editor-in-Chief for more than twenty years. Between 2014 and 2019, he was Editor-in-Chief of *Limnology & Oceanography*, the official journal of the American Society of Limnology & Oceanography. Since 2021, he has served as co-Editor-in-Chief of *OLAR*, the journal of Ocean-Land-Atmosphere Research, a new effort between the American Association for the Advancement of Science and a major new oceanographic lab in southern China.

AUDREY LEE, PHD

With decades of unique experience, Dr. (Ja-Chin) Audrey Lee is a respected energy policymaker, economist, and executive, and she is among the foremost distributed energy resource experts in the United States. She led a team that developed Sunrun's first, and probably America's first, residential virtual power plant projects (VPPs) with Laura Fedoruk and Steve Wheat at Sunrun. And she led the creation of four commercial-scale demand-response energy-storage projects for Advanced Microgrid Solutions. She also worked on energy policy as a Senior Economist for the Department of Energy's Office of Policy and International Affairs, executing global energy modelling and analyzing energy programs and standards for the White House; and for the International Energy Agency on the China section of its World Energy Outlook. Lee was later appointed Advisor to the President of the

California Public Utilities Commission, where she led development of the first-in-the-nation rules for energy data sharing. She holds a PhD in Electrical Engineering from Princeton University and is now Microsoft's Senior Director for Datacenter Energy Strategy, serves on the Clean Energy for America Education Fund, advises start-ups and investors, serves on the board of Redaptive, an energy efficiency-as-a-service company, and on the advisory board of CelerateX, an ESG investment platform. Her expertise is rare and highly prized as we dramatically ramp renewables to mainstream dominance.

LAURA FEDORUK, DATA SCIENTIST

Laura Fedoruk holds a bachelor's degree in Engineering Physics and a master's degree in Resource Management and Environmental Studies from the University of British Columbia and worked previously as a mechanical design engineer and project manager on net-zero-energy buildings and campuses such as Google's Bay View campus in Mountain View. She also led data science on Dr. Lee's team at Sunrun. Fedoruk developed the analyses and background information that convinced state regulators in the United States to open up wholesale market opportunities to distributed energy resources. These had previously been available only to large-scale power resources. She now works as Technical Program Manager on Project Tapestry at X, The Moonshot Factory (formerly Google X).

STEVE WHEAT, GRID PROGRAM MANAGEMENT

Steve Wheat was Senior Program Manager for Grid Services, working with Fedoruk and Dr. Lee when they pioneered the concept of virtual power plant (VPP) programs with client utilities such as National Grid in Massachusetts, Orange & Rockland in New York/New Jersey, and PSEG on Long Island. Wheat holds a master's degree in International Business from Concordia University in Irvine, California. He is now Vice-President of Program Management at Swell Energy in San Francisco, where he is leading some of the most ambitious, forward-thinking VPPs in the world, much as he did at Sunrun.

NANCY E. RYAN, PHD, EMOBILITY ADVISORS

Dr. Nancy E. Ryan has more than thirty years' experience in the electricity industry, served as Commissioner for the California Public Utilities Commission, Partner at Energy and Environmental Economics in San Francisco, and taught economics at UC Berkeley's Goldman School of Public Policy. She holds a PhD in Economics from UC Berkeley and Bachelor of Economics from Yale. She is an economist at eMobility Advisors focused on the low-carbon transformation, with an emphasis on transportation electrification and renewable energy. She has worked as a consultant, a regulator, an academic, and an environmental advocate specializing in the economics of renewable energy, the smart grid, electric transportation, and the implementation of cap and trade for the electricity sector.

PROFESSOR JEFF DAHN, DALHOUSIE UNIVERSITY

Professor Jeff Dahn is recognized as one of the pioneering developers of the lithium-ion battery that is now used worldwide in laptop computers, cellphones, cars, and many other mobile devices. Dahn and his research teams have worked closely with Tesla, NOVONIX, and other progressive firms. He is a Professor in the Department of Physics and Atmospheric Science and the Department of Chemistry at Dalhousie University and is invited as a keynote speaker to virtually all significant battery-technology conferences. He obtained his BSc in Physics from Dalhousie University in 1978 and his PhD from the University of British Columbia in 1982. From 1982 until 1985, he conducted research at the National Research Council of Canada before working at E-One Moli Energy until 1990, when he accepted a faculty position in the Physics Department at Simon Fraser University. He moved to Dalhousie in 1996 as NSERC/3M Canada Industrial Research Chair in Materials for Advanced Batteries, a position he held for the next twenty years. In June 2016, Dahn began a long-term partnership with Tesla Motors, in order to improve the energy density and lifetime of lithium-ion batteries, along with reducing their cost. He has received numerous awards, including the following: International Battery Materials Association (IBA) Research Award (1995); Herzberg Medal, Canadian Association of Physicists (1996); Fellow of the Royal Society of Canada (2001); Medal for Excellence in Teaching (2009) from the Canadian Association of Physicists; The Rio-Tinto Alcan Award from the Canadian Institute of Chemistry (2010); the ECS Battery Division Technology Award (2011); the Yeager award from the International Battery Materials Association (2016); the Inaugural Governor General's Innovation Award (2016); and the Gerhard Herzberg Canada Gold Medal for Science and Engineering (2017).

MICHAEL BARNARD, CLEANTECHNICA.COM AND FLIMAX

Michael Barnard is Founder and Chief Strategist of TFIE Strategy Inc. He provides future-oriented decarbonization technology and investment guidance to multi-billion-dollar industrial firms, infrastructure funds, and venture capitalists. His assessments of global grid storage, aviation, and maritime technology reform have gained significant attention in multiple industries. He publishes regularly in CleanTechnica.com and Illuminum.com, and his work has also appeared in *Newsweek*, *Forbes*, *New Atlas*, and *Quartz*. He hosts a regular decarbonization technology–oriented podcast with Redefining Energy, a clean energy platform that caters to the global investment community. Barnard is on the Advisory Board of electric aviation firm FLIMAX, has served on other battery and aviation startup advisory boards, and held the position Senior Fellow—Wind for the Energy and Policy Institute in Washington, DC. He co-founded Distnc Technologies, a healthy buildings design consultancy, and held senior systems architecture, transformation leadership, and sales roles with IBM on three continents.

DAVID BARNES, PHD, BRITISH ANTARCTIC SURVEY

Dr. David Barnes, of the British Antarctic Survey, is a marine ecologist who spends months and even years on research stations and vessels in the Arctic and West Antarctica. He is an advisor to the Intergovernmental Panel on Climate Change (IPCC); has been part of more than a dozen research expeditions, serving as leader on at least six of them; and was lead author on the majority of some 296 peer-reviewed publications, which have been cited more than twenty-four thousand times. He is a visiting lecturer at University of Cambridge, has taught at the University of Cork, in Ireland, has supervised fifteen successful PhDs, and is currently working with five others. Winner of a 2023 Polar Medal, he has also contributed as author/editor to several books and given more than seventy television, radio, and news interviews in fifteen countries, during which he has been quoted as saying, "We urgently need to help nature help us . . . nature-based solutions are the low-hanging fruit of climate solutions, but they should be carefully designed." He breaks down ways in which reforestation can be more effective and recommends approaches to planning for 30 by 30 (protection of 30 precent of land and sea by 2030) that could be more discriminating. He notes that protection of our marine environment lags behind protected area on land, with polar seas lagging behind the furthest, and reveals how the seafloor in polar regions (and particular habitats there, such as fjords) soaks up disproportionately high amounts of CO_2 during interglacial periods, such as in the present time.

KATE GAERTNER, TRIPLEWIN ADVISORY, PORTLAND OREGON

Kate Gaertner of TripleWin Advisory in Portland, Oregon, holds an MBA from the Wharton School at the University of Pennsylvania, an MSc in Sustainable Management from the University of Wisconsin, and is the CEO of TripleWin, which serves clients such as the Linux Foundation and StockX. She is an expert on progressive corporations, global food systems, circular economy, and greening company supply chains and maintains a strong network of corporate contacts around North America and the world. Gaertner has written for *Forbes*, *Fast Company*, *GreenBiz*, and is the author of *Planting a Seed: 3 Simple Steps to Sustainable Living*, which earned the 2022 Silver Nautilus Book Award and was voted one of the top twelve Global Sustainability Books to read in 2022. Before TripleWin, she worked for SiriusXM Satellite Radio, Ziff Davis Media, and Time Inc., and one of her first business ventures was an activewear company called OMALA, which designed and manufactured clothing made from sustainable fibers.

LUXMY BEGUM, PHD, PENG, PMP

A wastewater-treatment and resource-recovery specialist, Dr. Luxmy Begum, PEng, holds a PhD in Environmental Engineering from the University of Tokyo, a master's from AIT in Thailand, and a degree in civil engineering from the highest-ranked engineering institute in India (IIT, Madras). She has more than twenty years of professional experience in the water- and waste-management sector. This includes undertaking treatment plant upgrades, technology selection, evaluation, and procurement for water, wastewater, and anaerobic digestion plants. She manages EcoAmbassador. com, which provides extensive information, handbooks, and other training resources on water pollution causes and solutions, including cutting-edge knowledge on how medium-to-large wastewater-treatment plants can move toward energy efficiency and energy neutrality.

MARY D. NICHOLS

Mary D. Nichols is the former Chair of the California Air Resources Board, working with Governor Edmund G. Brown Jr. (1975–1982 and 2010–2018), Governor Arnold Schwarzenegger (2007–2010), and Governor Gavin Newsom (2019–2020). She led the crafting of California's internationally recognized climate action plan. She also served as California's Secretary for Natural Resources (1999–2003), appointed by Governor Gray Davis. She was senior staff attorney for the Natural Resources Defense Council, Assistant Administrator for US EPA's Office of Air and Radiation in the administration of President William Jefferson Clinton, and headed the Institute of Environment and Sustainability at UCLA. During her career, she has played a key role in progress toward healthy air in California and across the United States.

BF NAGY, AUTHOR/EDITOR

A prolific climate solutions writer and futurist, BF Nagy is among the world's leading experts on green buildings and communities, climate solutions, climate economics, technologies, and disinformation. He is also the author of *The Clean Energy Age* (Rowman & Littlefield 2018), has written more than two hundred feature magazine articles for the *New York Times, Corporate Knights, Globe & Mail, National Observer, Open Access Government*, and is a regular contributor to several others. He has visited green projects all over North America and beyond, documenting the struggles, secrets, and best practices of thousands of climate heroes who are accelerating change and winning against profit-taking fossil fuel companies. He is known for employing a plain-language, storytelling approach that makes it easy to understand complex global problems and opportunities for positive change. Nagy is also an award-winning playwright and popular speaker, produces clean energy films and videos, and hosts webinars for city planners on behalf of Canada's federal government. In addition to his profile in the green buildings and communities field, he studies electrified transportation, renewables, electricity storage, protecting water, natural refrigerants, artificial intelligence, cybersecurity, robotics, drones, virtual reality, modern media, and 3D printing.